THE EVOLUTION
of
Plants
and
Flowers

by Barry Thomas

THE EVOLUTION OF
Plants and Flowers

by Barry Thomas

With eight double-page paintings by Tony Swift

**St. Martin's Press
New York**

Stoneworts are living members of an ancient group
of plants which appeared 400 million years ago. They
live in fresh or brackish water and during life become
impregnated and encrusted with lime obtained from the
surrounding water. Although the plants themselves
decay when they die and sink to the bottom, the lime
skeletons survive as fossils, providing accurate
replicas of the plant's form.

For D

The author and
publishers would like to thank
Dr C. R. Hill and Dr Margaret
Collinson for all the help they have
given in the preparation of this book.

Printed in Spain

For further information write:
St. Martin's Press
175 Fifth Avenue
New York, New York 10010

ISBN 0-312-27271-5
Library of Congress Catalog Card Number 81-50779

Contents

Introduction

Vegetation clothes the world in many varied ways with a bewildering display of plants of all shapes and sizes. These living organisms produce the food upon which animals, including man, depend for survival. Equally, some plants depend on animals for successful pollination and seed dispersal. For over 4,000 million years an intricate balance has been maintained as more and more complex plants and animals have evolved.

Like all living organisms, plants ultimately die and the majority decay. Sometimes, however, they escape from final destruction and survive as fossils, entombed in rocks. It is these remains that allow us to piece together the scattered jigsaw of plant evolution.

We know that life began in the early oceans with minute single-celled organisms, sheltering there from the lethal effects of the primitive atmosphere. For thousands of millions of years plant life was confined to watery habitats but ultimately the time came when some plants moved onto dry land. This heralded a new era of plant evolution and diversification with one change following another much more rapidly. Plants increased in size, became more efficient at all their living processes and, most of all, modified the ways in which they reproduced, making further evolution more likely.

A thunderstorm over the Grand Canyon, Arizona. Flashes of lightning were probably very frequent in the atmosphere surrounding the Earth 4,000 million years ago. With the sun's radiation they provided energy to link groups of simple chemical substances together. These formed the basis for the first forms of life.

Understanding reproduction is the key to disentangling the jumble of living and extinct plants. Simple mechanisms gave way to more sophisticated ones that were better suited to dry land habitats. A dependence on wind instead of water and then on animals instead of wind led ultimately to the supreme group, the flowering plants. The increasingly complex inter-relationship of plants and animals speeded up the rate of evolution not only of the flowering plants but of insects, birds and mammals as well.

This accelerating series of changes led finally to the appearance of man, who is now seeking to manipulate plant evolution for his own needs. Plant breeding may encourage the changes needed to provide food for a growing world population but there is a danger that wholesale environmental destruction may eliminate the wild plants needed for research before they have been fully studied.

We can use our knowledge of the past evolutionary history of plants and flowers or we can ignore it: for a few more years the choice is still ours.

What is a plant?

Most people know what plants are and have some understanding of how they differ from animals. This knowledge is usually gained from the larger, more recognizable plants and animals we see in everyday life. Defining a plant accurately is much more difficult. There is such a range of living organisms that any straightforward distinction is impossible and the smallest and most simple living things are so difficult to classify that many have been described as both plants and animals.

One of the main characteristics used to distinguish between plants and animals is mobility: the vast majority of plants are stationary, while animals can generally move around. But there are important exceptions in both groups. Large numbers of microscopic algae, the group which includes seaweeds, swim in water and many land-dwelling plants produce freely moving sperm during sexual reproduction. In the animal kingdom many creatures – such as sponges and sea anemones – spend their whole adult lives attached firmly to rocks and stones.

The second main distinguishing feature between plants and animals is their nutrition. All living organisms, whether plants or animals, need energy which they obtain from water, mineral salts and easily broken down organic foodstuffs. Animals have to feed either on other animals or on plants to obtain this energy but green plants can literally live on air; in sunlight they are able to produce sugars from carbon dioxide in the atmosphere while giving off oxygen as a by-product. This process, called photosynthesis, uses energy from sunlight, trapped by the green pigment chlorophyll. Of course there are exceptions to this method of nutrition. Colourless plants, for example, cannot photosynthesize. Fungi and certain flowering plants are dependent on outside sources of food and live either as parasites on other living things or as saprophytes on dead and decaying organic matter. We cannot say that all plants photosynthesize but we *can* say that only plants do so.

The capacity of plants to produce oxygen is unique. Almost all living things need oxygen in order to respire and give off carbon dioxide as a waste product. Green plants are no exception to this, but they recycle the carbon dioxide into oxygen and sugars in photosynthesis. During daylight these plants therefore produce an excess of oxygen because the rate at which they manufacture it is quicker than the rate at which they use it up in respiration. As we shall see, this ability to produce oxygen is vital for both animal and plant evolution. Indeed, without oxygen-producing organisms, life on Earth as we know it could not exist.

The structure of a plant

The first plants to exist on Earth were single celled organisms which absorbed anything they needed from the environment through their cell wall. As many celled organisms evolved, individual cells became more and more specialized until eventually different parts of the plant developed different functions. The shapes and structures of plants today are related to their ways of life and the environments in which they must survive.

In its most complex form a plant consists of a rooting organ which holds it in the ground and absorbs water and mineral salts; and an aerial shoot of stems, leaves and reproductive organs, the flowers. In the interior layers of the leaves, the mesophyll, gases are exchanged and food is produced. The cells responsible for photosynthesis are surrounded by an outer protective layer, the epidermis, covered by a waterproof coat, the cuticle. Carbon dioxide and oxygen enter and leave the plant through special apertures in the epidermis called stomata, which can be opened or closed by a pair of guard cells. When the stomata are open, the air sur-

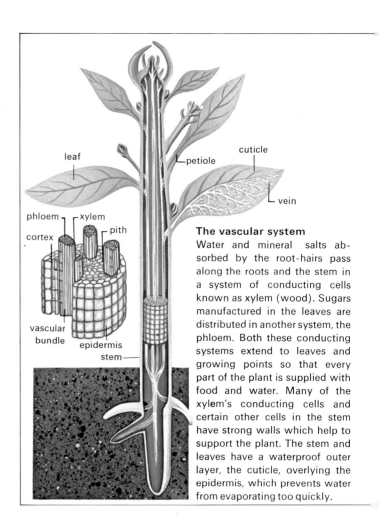

The vascular system
Water and mineral salts absorbed by the root-hairs pass along the roots and the stem in a system of conducting cells known as xylem (wood). Sugars manufactured in the leaves are distributed in another system, the phloem. Both these conducting systems extend to leaves and growing points so that every part of the plant is supplied with food and water. Many of the xylem's conducting cells and certain other cells in the stem have strong walls which help to support the plant. The stem and leaves have a waterproof outer layer, the cuticle, overlying the epidermis, which prevents water from evaporating too quickly.

flower

All land plants consist of some kind of anchoring system, a stem, leaves and a reproductive organ, here the flower. Through their delicate root hairs, roots absorb water and mineral salts from the soil. The leaves are the plant's food factory. They use light and carbon dioxide in a process called photosynthesis to make the sugars the plant needs for growth. As a by-product, oxygen is released. Carbon dioxide and oxygen enter and leave the plant through tiny pores, the stomata.

oxygen

sunlight

carbon dioxide

rounding the leaves and the air from the spaces between the cells inside mix. Carbon dioxide can thus be absorbed and the excess oxygen produced by photosynthesis can escape into the atmosphere.

It is also through the stomata that much of the plant's water is lost by evaporation. If a plant does not replace the water it loses, it wilts, so the tiny root-hairs of the actively growing roots continually absorb water and mineral salts from the soil. As a result, a stream of weak mineral solution constantly flows throughout the plant, carried in a system of woody tubes called the xylem. At the same time food reserves are carried away from where they are made in the leaves and distributed throughout the plant in another transport system, the phloem. The xylem and phloem systems are arranged in distinct strands called vascular bundles. Their number and the particular way they are arranged varies greatly from species to species and they are often used to distinguish between otherwise similar stems.

Plants naturally need to reproduce so that they may survive and spread. There are many varied methods by

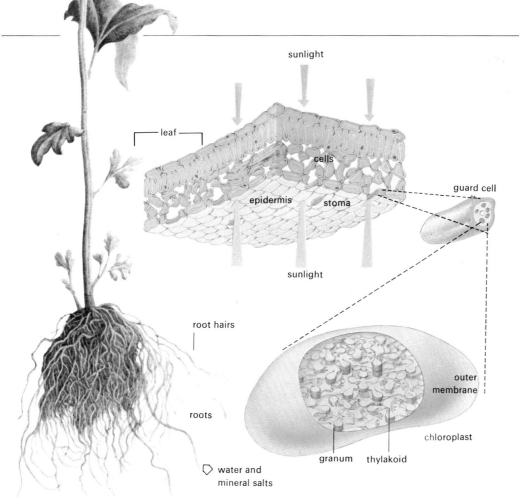

sunlight

leaf

cells

epidermis   stoma

guard cell

sunlight

root hairs

roots

granum   thylakoid

outer membrane

chloroplast

water and mineral salts

Photosynthesis
Photosynthesis can only take place in a plant which contains the green pigment chlorophyll. Chlorophyll traps sunlight and sunlight provides energy for the reactions that convert carbon dioxide and water to sugars. Chlorophyll is usually found in the guard cells which control the opening and closing of the stomata as well as in the inner leaf cells and sometimes also in the ordinary cells of the epidermis. It is contained in tiny structures called chloroplasts, only 4-6 microns across. (1 micron is 1 millionth of a metre.) Inside the chloroplasts are the thylakoids, arranged in stacks called grana. The actual chlorophyll pigment is on the walls of the thylakoids.

which they can do this with some being much more complicated or efficient than others. Perhaps the best known is the formation of seeds but as we shall see, seeds are really the end product of a long series of evolutionary changes that took many millions of years to complete.

How plants are classified

Plants are very varied in appearance and have a be-wildering number of names. There are, of course, underlying similarities and plants with most features in common can be formed into groups which are then joined into larger and larger units, all members of which have certain features in common. The classification of plants depends mainly on the different methods they use to reproduce and, to a lesser extent, on their structural, genetical and biochemical features. Some classification systems emphasize one aspect more than another and produce different groupings. For example, two plants that have similar structures may be grouped together in a system that takes structures as the most important distinguishing feature but separated in a system which relies more on their biochemistry. It is almost impossible for everyone to agree on the best system to use.

Did evolution take place?

If we look at today's plants, it is obvious that some are much more complex in structure than others and may have more sophisticated reproductive organs. The theory of evolution is that these complicated plants have developed from much simpler forms by a long series of changes taking place over thousands of millions of years.

Changes may have been chance mutations, shuffling of DNA or possibly adaptations to a changing environment. Some of the changes must have been less successful than others, giving plants no advantage – or even putting them at a disadvantage. These plants would have become part of a dead end line, leading to eventual extinction. If an evolutionary change was successful, however, the earlier parent type of plant sometimes disappeared because it could not compete with the new forms. These, being better suited to their environments, could take over existing habitats or even invade new ones. In other cases very simple forms have survived almost unchanged to the present day.

The simplest plants today are bacteria and tiny, water-living forms, the single celled algae. Plants such as ferns, which produce spores instead of seeds and flowerless seed producers such as the conifers, are

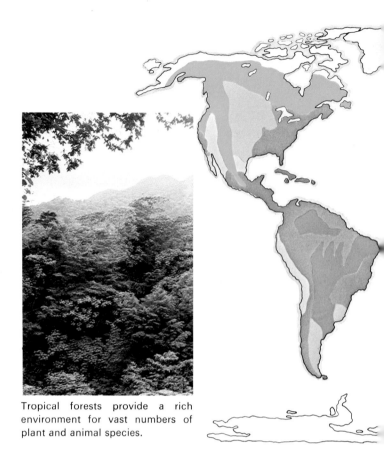

Tropical forests provide a rich environment for vast numbers of plant and animal species.

Tropical savannahs are grasslands, often dotted with acacia and other thorn bushes.

intermediate. Flowering plants are the most advanced.

The evidence in the rocks supports the general idea of plant evolution although it never answers all the questions. Rocks can now be reliably dated and we can obtain an idea of when the various plant groups first appeared by studying their fossilized remains in rocks of different ages. They show clearly that as the rocks get younger, the plant fossils they contain become increasingly complex. Unfortunately we cannot yet link up all the major groups of plants into an evolutionary series because many intermediate stages are just not known. Instead of being a single chain of events, plant

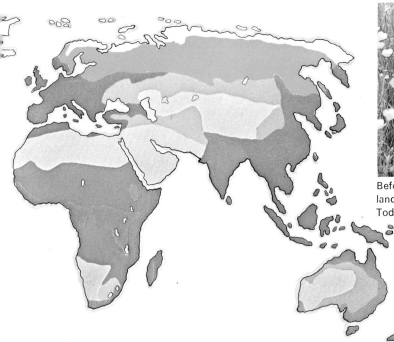

Before the use of weedkillers became widespread, temperate grasslands were a mixture of grasses and colourful herbaceous plants. Today many of the wild flowers have disappeared.

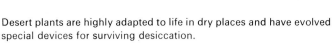

☐ tundra	grassland	coniferous forest
desert and semi-desert	savannah	other forest and woodland

The main vegetation zones today. Plants have adapted to most types of climate and habitat, even to the extremes of arid deserts and of cold, high mountains. The greatest variety of species is found in tropical forests, where conditions are ideal for rapid growth. North and south of the equator, the variety gradually becomes less until in the extreme north and south only a handful of species survives.

Conifer forests dominate 8% of the land's surface, mainly in the cooler, northerly areas.

Deciduous forests once covered much of Europe but now survive only in isolated patches.

Desert plants are highly adapted to life in dry places and have evolved special devices for surviving desiccation.

evolution shows a network of different changes, slowly separating out the major groups and splitting them up into recognizable units.

In looking at plant evolution as a whole we must be aware that there are varying patterns of change. New forms may appear abruptly by mutation or by a gradual series of small modifications and ancient forms may remain stable for millions of years. We know that the processes of evolution are tied up with the genetical make-up of the plant and we can often prove that they have taken place. But there is no way that we can ever be absolutely certain of how or why changes occurred.

Fossils: reading the rocks

Some plants can live for a very long time, like the 5,000 year old bristle-cone pines, or can grow to enormous heights, like the 100m tall Californian redwoods. Ultimately, of course, all living things die and from that moment microbes steadily break down the remains, eventually destroying them completely. Long lived plants also shed parts of their structure. Leaves, flowers and often even branches and bark drop and decay.

Although plant parts disappear, their decay recycles vital elements back into the environment, making it possible for more plants to grow in their place. As the dead parts are broken down carbon is released as carbon dioxide into the atmosphere and nitrogen as ammonium compounds into the soil. These and other essential nutrients can then be reabsorbed and used by other plants. In this way the natural processes of decay actually ensure that the supply of plant nutrients will continue.

The rate of decay varies according to the type of tissue because some cells are more resistant than others to the attacks of microbes. Plant cells have relatively firm walls and are less likely to decay than the cells in the softer parts of animals. The chemicals of which the walls are made up control the rates at which they decay. The basic material of all cell walls is a carbohydrate called cellulose but there may also be thickening, supporting substances such as lignin (in wood), suberin (in cork), cutin (in surface layers of stems and leaves) and sporopollenin (in the walls of spores and pollen grains). All these help to resist decay with cutin and sporopollenin being the most effective.

Sometimes plant fragments may be buried in environments where decay cannot occur. Acidity, lack of oxygen and mineral-rich water can all prevent microbes from breaking down the tissues. Thus if conditions are right, plant remains can be preserved and ultimately fossilized.

All remains, whether of plants or animals, must be permanently entombed in rocks if they are to survive for any length of time. For this to happen they must be covered by sediments – sand, silt or mud – as or soon after they fall. As more and more layers of sediment accumulate, the grains of sand or mud are pressed together into a solid rock which seals in the plant or animal remains and fossilizes them.

How fossils are formed

There are many different types of fossils, formed in different ways. Usually the slow hardening of the rock flattens plants into black coaly fossils called compressions. When the rock is split open the compression stays

In the Triassic period this desert in Arizona was covered by conifer forests with some cycads and ferns. Volcanoes were active in the area and sometimes trees fell into water that was hot and contained silica. When this happened, the trunks became impregnated with silica and, buried in the mud, were preserved as fossils. Later the land was raised by earth movements and as the rocks were worn down by wind and rain, the harder, petrified trunks, were exposed.

This fossil leaf must have been buried very quickly in fine grained mud and so protected almost completely from decay. Even its original green colour has been preserved. It was found in Germany, in rocks from the mid-Eocene period, so is between 40 and 50 million years old.

Different parts of a tree have different chances of being fossilized. Although trunks are one of the most durable living parts they are often the rarest fossils for each tree has only one trunk. Branches, twigs and leaves are usually more numerous, followed by pollen, seeds and flowers. Flowers, however, are usually soft and short-lived and are rarely preserved. The most common plant fossils are pollen grains which are produced in enormous numbers.

Amber is fossilized resin, the sticky gum that oozes out when trunks or branches of conifers and some flowering plants are damaged. Some 20 million years ago this catkin was trapped in a drop of resin which hardened around it, preserving it from decay.

attached to one surface, leaving an imprint on the other. The fossil may be affected by various types of geological activity while it is still inside the rock. Water or air may penetrate or there may be further flattening. Sometimes the coaly compression is completely destroyed, leaving only imprints. Both types of imprints are called impression fossils.

A kind of three-dimensional impression may also be found if hollow plant parts are filled in either before they are buried or while sediments are settling over them. Such fossils are called casts although they may also have a carbon compression attached to them. Sometimes the remains are filled early on with concentrated mineral solution and are very well preserved. These fossils, called petrifactions, can be cut open to show internal structures and details of the cells. This is possible because minerals have crystallized within the cells, replacing the pattern of each part. Fine details such as starch grains and even the tiny nuclei of individual cells have been preserved in this way.

Coal seams are known in rocks of varying ages, although the most economically important come from the Carboniferous period (345-280 million years ago). Coal is made up of millions of compression fossils of plants which once grew together in bogs and swamps. These swamps formed in large, waterlogged basins which often covered hundreds of square kilometres. Trees and other plants grew either on islands of dry

How fossils are formed

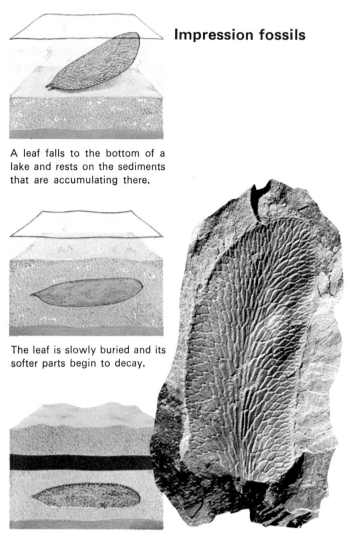

Impression fossils

A leaf falls to the bottom of a lake and rests on the sediments that are accumulating there.

The leaf is slowly buried and its softer parts begin to decay.

The leaf is flattened by the accumulating sediments, which harden into rock. If air or water seep around it, the actual leaf decays completely, leaving only an impression in the rock.

Impression fossil of *Linopteris*, the leaf of a seed fern from the Carboniferous period.

Compression fossils

A leaf falls to the bottom of a lake and rests on the sediments that are accumulating there.

The leaf is buried very quickly, before it has started to decay.

The plant and the sediments are pressed and flattened by the weight of new sand and mud particles but the plant does not decay. When the rock is split open, its hardened remains are still visible.

Compression fossil of *Neuropteris*, the leaf of a seed fern from the Carboniferous period.

land or in the shallows. Layers of peat, made up of partly rotted plant debris, accumulated in the stagnant water, becoming as much as 30m thick. The land was sinking very slowly and over millions of years plants took root, matured, died and were replaced. Occasionally, however, the land subsided more rapidly and water flooded in, submerging or knocking over the trees. Grains of sand and mud carried along by the water built up layers of sediment over the submerged, dying vegetation and peat, which then began its gradual hardening and conversion into coal. The first rush of sediments also trapped many plant fragments and these are often found as compression fossils in the rocks which roof the coal seams.

After a time the rapid earth movements would cease and the swamp land would begin to sink more slowly again. Plants would once more become established there – only to be submerged once more if rapid subsidence occurred again. In many areas where coal is found the layers of fossil fuel are divided by thick layers of sediments, showing that swamp vegetation grew and was destroyed over and over again as the underlying land subsided and rose.

Sometimes the land would sink so far that sea flooded over it, producing a different type of sediment called a marine band. This happened only rarely but the evidence left behind is of immense value to the study of fossil plants. Minerals in the sea water seeped down

Petrifactions

When a tree trunk or stem falls to the bottom of a lake, its soft outer covering and soft centre rot away.

A cone falls into a mineral solution such as lime, sulphur or a volcanic spring.

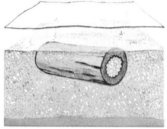

The mineral solution seeps into the cone and crystallizes in its cells, replacing the pattern of each part.

If the trunk is buried slowly, the outside rots away but the hard wood remains and the centre is filled in, forming a cast.

Cast fossils

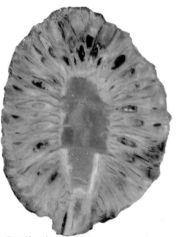

When the sediments are compressed the hard, mineralized cone keeps its original shape. When it is split open, even fine details of its original structure can be seen.

Petrifaction of the cone of a monkey puzzle, *Araucaria mirabilis*, which grew in South America during the Jurassic period.

Pressure from the accumulating sediments flattens the wood and the cast slightly. Because there is an even pressure from each side, the width remains the same.

Cast fossil of the central pith cavity of the stem of *Calamites*, a Carboniferous tree.

through the peat to impregnate and petrify large lumps of it. These, known as coal balls, can be cut open and tell us a great deal about the types of plants which formed the swamp vegetation millions of years ago.

Naming and understanding fossil plants

Perhaps the greatest problem facing any palaeobotanist is trying to put together the various bits and pieces of fossils into whole plants. Plants shed parts of themselves during life and fall apart when they die so any collection of fossils from one place will almost certainly contain the fossilized parts of many plants but, equally likely, it will not contain all the remains of any complete plants. For this reason a unique system of naming separate fossilized plant organs is necessary to cope with the problem of comparing lists of fossils from different places.

In the eighteenth century the Swedish botanist Carl Linnaeus devised a system of naming all living things. Now called the Linnaean system, it provides for what we call generic and specific names which are often in latin or have latinized endings. The meadow buttercup, for example, is correctly called *Ranunculus acris*, the creeping buttercup is *Ranunculus repens* and the bulbous buttercup is *Ranunculus bulbosus*. All three, and a great many more possess certain very close similarities and belong to the same genus, *Ranunculus*. The second part of the name is the specific name and shows that they differ from each other in detail.

Another genus which also includes several species is *Thalictrum*, the meadow rues. *Thalictrum* and *Ranunculus* are not similar enough to belong to the same genus but they still share certain characteristics and are said to belong to the same family. There is therefore a hierarchical system of classification in which species belong to genera, genera to families, families to orders, orders to classes and so on up to the division. At the very top of the scale is the plant kingdom, which includes all living things not classed as animals, from microscopic algae to giant trees.

To cope with fossil plants, this system has been extended to separate organs. There are genera and species of leaves, stems, roots, reproductive organs and even isolated pollen grains and spores. The various fossilized parts of a large plant may therefore have many different names, just as if they were separate plants themselves. It is a system which enables scientists to produce lists of fossils for comparison but it does not help with the reconstruction of whole plants.

Separated plant organs are put together for a variety of reasons: they may be consistently found together; they may look alike or their internal anatomy may be similar; they may have distinctive hairs or scales. But all these reasons merely allow us to guess what the original plant was like. However valid the reasons seem, and however often they are repeated, it is still a guess. The ultimate proof rests on the chance finding of connected fragments. Many reconstructions are, therefore, the result of intelligent interpretation of evidence plus a little guesswork. This is why many scientific reconstructions appear a little vague in parts. A plant may be shown without the base of its stem and roots because these have never been found as fossils; or its reproductive organs may be quite unknown although other parts found together can be shown complete. In some cases it is even difficult to be sure how big the plant was in life. We do have a good idea of what many extinct plants looked like but equally we know very little about a great many more.

Dating rocks and fossils

The other major problem in studying fossil plants is the need to date collections accurately and to work out the relative ages of the various plants. Only when this is known is it possible to determine when the plants changed during their evolution, as well as the ways in which they have altered.

Geologists date rocks by a variety of methods that allow them to arrange the various rock layers or strata in a series based on their relative ages. The familiar

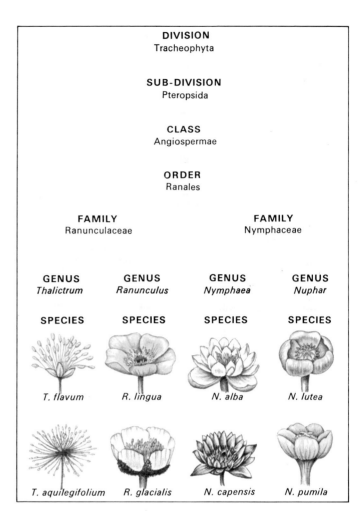

Classification is based upon similarities and differences between plants and attempts to arrange them into groups organized in hierarchical order. The species, at the base of the hierarchy, is a group of plants so similar that they can interbreed freely. They are arranged into genera, families, orders, classes, sub-divisions and divisions with the number of similarities becoming fewer at each level.

geological column, covering nearly 600 million years, was originally compiled in this way. Much of the relative dating is based on accurate fieldwork which has built up a knowledge of rock sequences that can be traced over wide areas and matched with similar sequences in other places.

The geologist begins by examining the strata to decide on the type of rocks, their relative thicknesses and the order in which they have been laid down. He then examines the minerals of which they are made up and deduces the conditions under which they were formed. Any fossils they contain must be carefully labelled with precise information about where they were collected for only in this way can we be sure of the facts about them. They can then be compared with plant fossils from other rocks of different ages. It is by such studies that the changing history of plant life has been gradually clarified.

Once the sequence is known, fossils can then be used

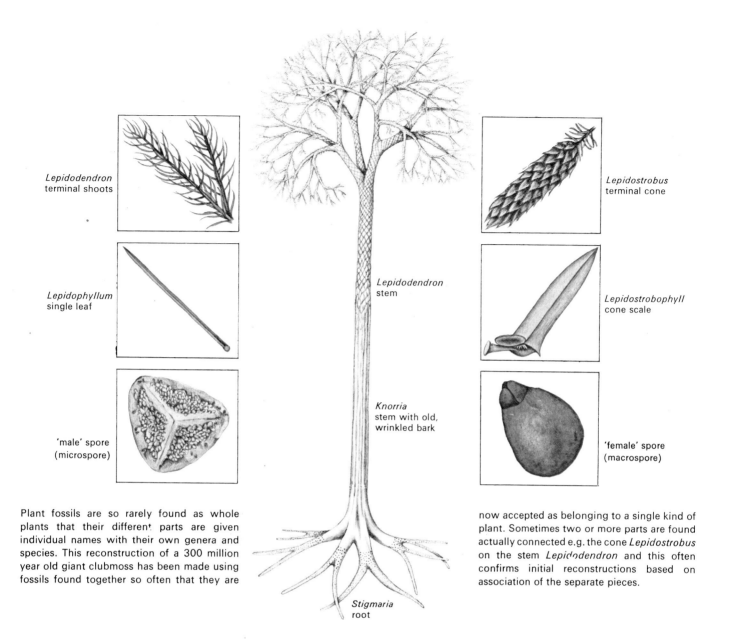

Lepidodendron
terminal shoots

Lepidophyllum
single leaf

'male' spore
(microspore)

Lepidodendron
stem

Knorria
stem with old,
wrinkled bark

Stigmaria
root

Lepidostrobus
terminal cone

Lepidostrobophyll
cone scale

'female' spore
(macrospore)

Plant fossils are so rarely found as whole plants that their different parts are given individual names with their own genera and species. This reconstruction of a 300 million year old giant clubmoss has been made using fossils found together so often that they are now accepted as belonging to a single kind of plant. Sometimes two or more parts are found actually connected e.g. the cone *Lepidostrobus* on the stem *Lepidodendron* and this often confirms initial reconstructions based on association of the separate pieces.

to date newly studied rocks. Fossils in rocks which are geographically very widely separated from the better known areas can be compared with fossils of known date found elsewhere. If they correspond, they are likely to be of the same age and so, therefore, are the rocks in which they occur.

Early in this century radioactivity began to be used to date rocks. Radioactive minerals occur in rocks from the Earth's inner core which have reached the surface, often erupting as volcanic lava. The radioactive element uranium 235 decays at a rate that can be measured and when it decays it changes into a special type of lead. By measuring the amount of lead and comparing it with the amount of uranium left in the rock, it is possible to calculate the length of time that has passed since the rock was formed. Other radioactive elements give even more accurate dates and using these, the length in years of all the geological periods has been carefully calculated and is generally accepted.

Reconstructing the landscape

More emphasis is now being placed on interpreting the environment in which plants grew, on placing them together in their ancient prehistoric landscapes. The rocks themselves tell us what the environment was like when they were laid down as sediments. It is possible to tell whether the layers formed above or below water, in shallow or deep water, even whether the water was fresh, brackish or marine. We know that in the Carboniferous period (345 to 280 million years ago) most plants grew in vast swamps, that in the Jurassic (190 to 136 million years ago) parts of Yorkshire were a large river delta and that in the Tertiary (64 to 2 million years ago) Oregon USA was an upland area, with plant remains preserved in small inland lakes.

It is important to know how far sediments and plant parts have travelled before being deposited and this, too, can often be deduced from rock types. If the sediments were laid down close to an area where plants were growing the fossils will represent these local plants. But if they have travelled a long distance, carried by fast flowing water, plants from widely separated habitats collected on the way may be fossilized together. To appreciate this, imagine the difference between plant debris falling into a small lake in the middle of a forest and that carried along by a large river passing through a varied countryside.

The analogy of the lake and river highlights a further problem, the preferential sorting of plant fragments. Wood, twigs, fruits and seeds are more buoyant than leaves and will stay afloat longer. Thus any large river will carry some floating debris far out into its estuary or even into the sea, while it will drop waterlogged parts as soon as its current slows. The Tertiary plant remains in south-eastern England excellently illustrate this sequence of events, known as preferential sorting. It has resulted in the accumulation of leaves in Reading Beds in Berkshire, where sediments were deposited in the channels of a river flood plain. However, on sediments deposited further out to sea, for example the London clay of Sheppey, Kent, fruits, seeds and twigs are found. Ideally deposits from many environments and from the same time should be studied but this is rarely possible, either because the sediments are not exposed or because they have been eroded away. Much of London's underground railway and the older Thames tunnels are constructed through London clay sediments but only a small part of these rocks outcrop.

Fossil pollen

The study of pollen grains and spores – known as palynology – is a relatively recent field of research which really began in the 1930s. It has dramatically expanded in recent years with the growth of the oil industry and the need for quick and reliable rock dating. Plants produce enormous quantities of pollen grains or spores which can be carried for vast distances by the wind. Any area where sediments are being deposited will therefore be receiving large numbers of grains which become incorporated into sediments. Their resistant sporopollenin coats ensure that they survive the rigours of transportation and sedimentation, making them ideal subjects for fossilization.

As the world's vegetation changed so did the pollen and spore rain that was falling and becoming fossilized. Any sedimentary rock will therefore contain a selection of grains that should be very similar to collections from other rocks of the same age in the same general area. If the selection of pollen is different, the ages of the rocks

Preferential sorting
Leaves, twigs, fruits and seeds from the same tree fall into a swift-flowing river.

At first they are carried in the current but as the river slows and broadens, the leaves become waterlogged and sink.

Pieces of wood, fruits and seeds may float far out into the estuary, even out to sea, before sinking and being buried in mud.

Later, the deposits of the river or sea bed may be raised by earth movements, leaving the fossil-bearing rocks exposed.

are not the same. The methods of extraction are simple, consisting basically of chemically and mechanically breaking down the rock to release the grains. These can then be made transparent and stained for examination under a microscope.

A method like this allows fossils to be obtained from small samples of rock and is ideal for exploratory bore hole work. The rock samples taken from the bore hole cores can be analysed and dated and results are made available within a reasonably short time. Speed is obviously of paramount importance to an industry which is spending enormous amounts of money operating its drilling rigs.

Botanical, as well as geological, information can be gleaned from these core samples. The grains give some idea of the plants that were living in the general area, even if no other remains are found there. This assumes that we already know which plant produced the particular grain in the first place and in turn means that a reproductive organ full of grains must have been found before with a parent plant. This naturally applies to very few of the thousands of pollen grains and spores that are recovered in palynological investigations. But they are often a very great help in interpreting extinct floras and without such information our knowledge would undoubtedly be much poorer.

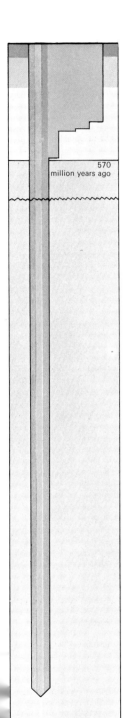

570 million years ago

4,500 million years ago

Left: The most spectacular diversification of plant forms has occurred in the last 600 million years. For the greater part of the Earth's 4,500 million year old history the plants were minute bacteria, various types of primitive algae and possibly also simple fungi.

Bacteria	Fungi	Blue-green algae	Mosses and liverworts	Algae	Clubmosses	Cycads	Ginkgos	Conifers	Angiosperms	Gnetales	Ferns	Horsetails			

Cycadeoids

Caytonia

Cordaites

Seed ferns

Trimerophyton

Cooksonia

2	TERTIARY	CAENO-ZOIC
64		
136	CRETACEOUS	MESOZOIC
190	JURAS-SIC	
225	TRIAS-SIC	
280	PERMIAN	PALAEOZOIC
345	CARBONI-FEROUS	
410	DEVONIAN	
440	SILUR-IAN	
530	ORDOVICIAN	
570	CAMB-RIAN	
	PRE-CAMBRIAN	PROTEROZOIC

QUATERNARY

million years ago

Plant evolution: the last 600 million years

Processes of evolution and extinction have resulted in today's main plant groups. Some evolved thousands of millions of years ago and remain almost unchanged. Others reached a peak in the variety of different species and then declined or even completely died out, long before man appeared. Today two groups dominate the land: conifers cover over 10 million square kilometres, mainly in the cooler, drier areas, while everywhere there are flowering plants. In the seas the algae have evolved in their own way, adapted to life in water.

23

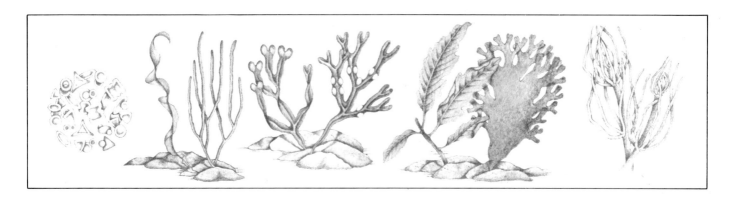

The first plants

It is generally assumed that the Earth was formed about 4,500 million years ago from a cloud of cold gases and dust rotating around the sun. Gases and dust particles were gradually drawn together and the accumulating mass began to spin on its own axis, drawing more and more particles together by gravitational force. The particles were packed more and more closely together until they formed a dense, almost spherical mass. At this stage there was no atmosphere and no water: life as we know it could not have existed.

Inside the newly formed planet were radioactive substances and these, with the pressure of gravity, caused the core to heat up. Hot gases and lava bubbled through the crust and the first primitive atmosphere formed – probably a mixture of carbon dioxide and water with minor amounts of methane and ammonia. At the surface the lava cooled and hardened. The water vapour condensed into water and, carrying other dissolved elements and compounds, settled on the surface as oceans, lakes and rivers.

Life is believed to have begun when groups of simple chemical compounds – sugars, phosphates and nitrogen – came together to form nucleic acids. The energy necessary to make this happen came from either the sun's radiation or lightning discharges, which were probably very frequent in the simple atmosphere of the time. However, although solar radiation may have provided the initial energy, it was almost certainly lethal to life. Deadly ultraviolet rays reached the Earth virtually unhindered and could even penetrate the surface waters to a depth of about ten metres.

It seems likely that the first steps towards life took place at the bottom of deep, stagnant pools. Here there was no danger of turbulence bringing the tiny forms within reach of ultraviolet radiation. These primitive forms of life fed on simple substances dissolved in the water and obtained their energy by fermentation processes which did not require free oxygen. Carbon dioxide was given off during the fermentation and this gradually altered the concentrations of gases dissolved in the water and eventually in the atmosphere itself.

As the amounts of gases in water and atmosphere changed, new organisms were able to develop which could produce their own food by photosynthesis. Three thousand million years ago, in the middle of the Pre-Cambrian period, the first bacteria, blue-green algae and possibly the green algae evolved and started the second change in the Earth's atmosphere. These first primitive plants were microscopic organisms consisting of only a single cell. Later they evolved into organisms made up of groups or chains of identical cells. Photosynthesis carried out by these simple green plants built up the amount of free oxygen in the atmosphere until it eventually reached a level of 0.2%. This was enough to screen ultraviolet light from all but the top few centimetres of water, allowing floating and shallow-water living algae to evolve.

The first fossil plants

The very oldest fossil plants are found in fine-grained rock called chert from the Pre-Cambrian period. Structures which look like bacteria and single celled, spherical blue-green algae have been found in the 3,000 million year old Fig Tree chert in South Africa. Similar forms and others which resembled the thread-like, filamentous kinds of blue-green algae have been found in 2,000 million year old Gunflint chert in Ontario, Canada and in the 1,000 million year old Bitter Springs chert in central Australia. In Australia there are also fossils of ancient aquatic fungi. It used to be thought that single celled green algae (the first plants with nuclei) also came from the Bitter Springs chert but it is now known that they were badly preserved fossils of blue-green algae.

The time of the first appearance of green algae is therefore unknown but there is circumstantial if not direct evidence that they evolved very early in the Earth's history. The evidence lies in hard chalky or siliceous knobs called stromatolites which are today produced by densely interwoven mats of blue-green and green algae. The oldest known stromatolites come from the Bulawayo limestone in Rhodesia and are 2,700 million years old. If these were formed by the same types of algal colonies existing today it would mean that green algae were already present nearly 3,000 million years ago, almost as early as bacteria and blue-green algae.

Stromatolites became much more widespread in the Cambrian (570 to 530 million years ago) when there seems to have been a natural balance between the algae's growth rate and the simple, single celled animals that ate them. As the smaller algae were eaten, larger and tougher types which were less suitable as food evolved and these produced a variety of stromatolite shapes. Other larger animals, similar to living marine worms, appeared in turn to make use of the new plants and as a result further changes occurred. In the end, however, the algae could not compete successfully and, by late Cambrian times, about 550 million years ago, there were far fewer of the large stromatolites. By the Carboniferous period, 345 million years ago, they had

almost disappeared. Today stromatolite producing algae live mainly in Australia and the Bahamas, surviving in specialized environments where there is little competition, in salt marshes and in places where strong currents or high sedimentation rates prevent animals from feeding on them continuously.

New forms of algae

During the Cambrian period other, larger green algae with several cells evolved. These had recognizable plant-like forms and their cells were no longer identical but had specialized functions. Some were so much like living seaweeds that they are included in the same families. Red and brown algae were also making their first appearance at about this time.

In the next periods, the Ordovician and the Silurian,

The evolution of the atmosphere
The first primitive atmosphere was formed by condensation and volcanic activity as the Earth's crust cooled and hardened. Consisting of water vapour, methane, ammonia and hydrogen, it allowed deadly ultraviolet rays to reach the Earth's surface unhindered.

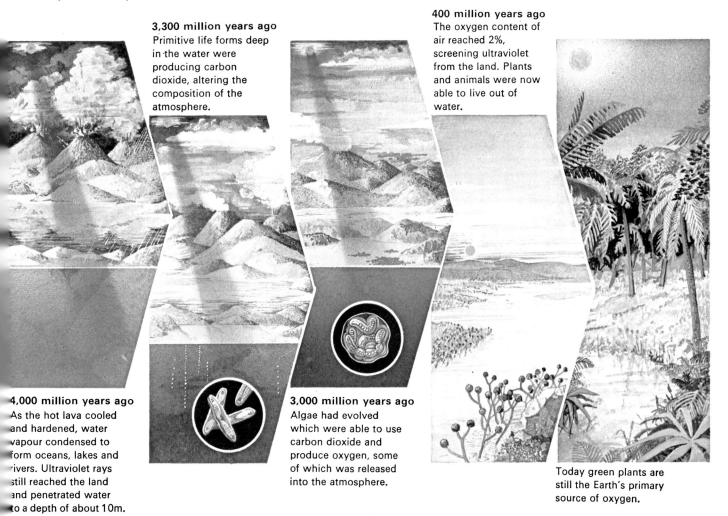

3,300 million years ago
Primitive life forms deep in the water were producing carbon dioxide, altering the composition of the atmosphere.

400 million years ago
The oxygen content of air reached 2%, screening ultraviolet from the land. Plants and animals were now able to live out of water.

4,000 million years ago
As the hot lava cooled and hardened, water vapour condensed to form oceans, lakes and rivers. Ultraviolet rays still reached the land and penetrated water to a depth of about 10m.

3,000 million years ago
Algae had evolved which were able to use carbon dioxide and produce oxygen, some of which was released into the atmosphere.

Today green plants are still the Earth's primary source of oxygen.

all the groups of algae except those that produced stromatolites increased, especially the reds and the greens. The forerunners of modern coral-building red algae are first found in the Ordovician period (530 to 440 million years ago).

In the Devonian period, which began some 400 million years ago, there were dramatic changes in the evolution of the algae. The older forms gradually disappeared and new kinds evolved. Some of these were to survive relatively unchanged to the present day, others rapidly died out. Blue-green filamentous algae were at their most abundant in this period and there are many almost identical forms still living. The stoneworts also appeared and they, too, have survived almost unchanged. They have many similarities with the green algae although they look rather different and live in brackish or fresh water. Their main stems carry whorls of side branches which occasionally include their distinctive sex organs. These, called oogonia, have a single egg cell surrounded by twisted spirals of other cells. Because they become encrusted with limestone they are often fossilized and are easy to recognize.

Some of the algae that evolved during the late Silurian and early Devonian periods were quite unlike any that occurred before or after that time. They might have been forms that were living either partly or entirely out of water – the first experimental land plants.

Life cycles: the algae

One of the reasons why these early water plants were able to evolve and change is to be found in their method of reproduction.

Every living organism contains in its cells' nuclei a set of chromosomes. These tiny bodies carry the genes which determine the special make-up of every individual. When a plant or animal reproduces, it passes its set of chromosomes and therefore its special characteristics on to the next generation.

The very simplest plants and animals reproduce by division. Each generation is the same as the one before as it contains an exact copy of the same set of chromosomes. The most complex organisms reproduce sexually, that is cells from two separate individuals must combine to produce a new generation. The offspring contains two sets of chromosomes, one from each parent. This means that characteristics are combined in different ways and makes evolutionary change possible.

In its life cycle an advanced alga shows a repeated pattern of non-sexual (better called asexual) and sexual reproduction called alternation of generations. At the asexual stage each plant contains two sets of chromo-

Trichodesmium, a primitive single-celled blue-green alga, similar to those that evolved in the seas 3,000 million years ago.

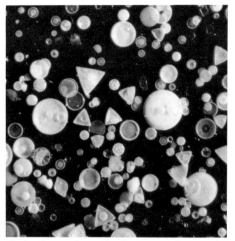

Skeletal remains of diatoms, microscopic algae which have their cell walls impregnated in life with silica.

Laminaria, a large brown alga frequently washed up on seashores all round the world.

26

A kelp forest off the coast of Australia.
Algae in general have no strong supporting
cells and are held upright by the water. Giant
Pacific kelps may grow up to 30m tall.

Right: A brown alga (*Fucus*) and the green
Enteromorpha (right).

Corallina officinalis, a red coralline alga. Red
algae usually grow in deeper water than
greens and browns.

somes, one from each parent: it is called diploid. When mature, the diploid plant produces reproductive cells of its own but by a special type of nuclear division called meiosis, these have only a single set of chromosomes. The reproductive cells, called spores, grow to adult form by themselves, asexually. The new generation of plants, each with only one set of chromosomes, are called haploid. They reproduce sexually.

Each haploid plant produces tiny reproductive cells called gametes, some of which are male (sperm) and some female (eggs). Each gamete contains its individual set of chromosomes. Male and female gametes are attracted to one another chemically and fuse together. Their chromosomes merge and produce an entirely new form called a zygote which now contains two sets of chromosomes, one from each 'parent' gamete. It is therefore diploid. The single celled zygote grows into another multicellular plant, the new diploid plant which produces spores so that the asexual part of the cycle can begin again.

This basic cycle has many variations and occurs in both single and many celled species. In multicellular plants the diploid stage which produces spores for asexual reproduction is called the sporophyte. The haploid stage which produces the gametes in the sexual part of the cycle is the gametophyte. Some plants look the same whether they belong to the diploid or the haploid generation but others are very different.

The gametes involved in fusion in the sexual part of the cycle may also vary. They are identical in the simplest forms but in more complex algae one, the female, may be larger and less mobile than the male. The female is sometimes completely immobile and, in an extreme case such as *Oedogonium*, is not only immobile but is also retained within the parent plant. The egg is fertilized within the gametophyte and the resulting zygote begins its growth there. When the female gamete is not mobile, the sperm, attracted by chemical secretions, swim towards it.

It is tempting to think of one generation in this complex life cycle as being the main 'adult' and the other as an intermediate stage. This is not correct: both the diploid sporophyte and the haploid gametophyte are fully mature plants, capable of reproducing. Both are vital to the survival of the species.

Although this type of reproductive cycle made major evolutionary changes possible, not all the ancient groups of algae attempted to colonize the land. Many continued to live in the sea, some remaining almost unchanged, others developing different life cycles and evolving into the great variety of forms alive today.

Algae were the first plants other than bacteria to evolve sexual reproduction in which the characteristics of two individuals are mixed and passed on to their offspring. They often have a complex life cycle involving both a sexual (gametophyte) and an asexual (sporophyte) stage. Sporophytes reproduce by releasing asexual spores that grow into male and female gametophyte plants. These in turn produce sexual spores, the male and female gametes — the plant equivalent of sperm and eggs. The gametes of opposite sex are attracted to one another and fuse, then grow to produce the next sporophyte generation. Sporophytes are diploid, meaning that they contain two sets of chromosomes, one from each parent gamete. Gametophytes are haploid, inheriting only one set of chromosomes from their parent.

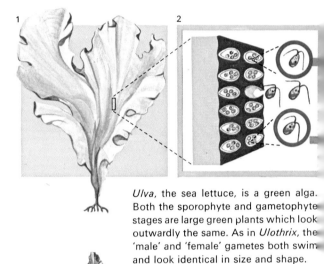

Ulva, the sea lettuce, is a green alga. Both the sporophyte and gametophyte stages are large green plants which look outwardly the same. As in *Ulothrix*, the 'male' and 'female' gametes both swim and look identical in size and shape.

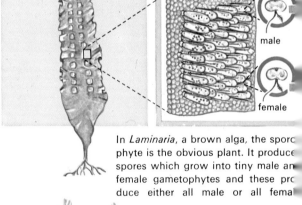

In *Laminaria*, a brown alga, the sporophyte is the obvious plant. It produces spores which grow into tiny male and female gametophytes and these produce either all male or all female

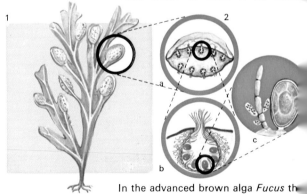

In the advanced brown alga *Fucus* the gametophyte stage has practically disappeared and is represented only by the sex cells themselves.

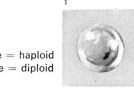

dark blue = haploid
light blue = diploid

Ulothrix has the most primitive type of life cycle. There are sporophyte and gametophyte generations but only the gametophytes are multicellular plants.

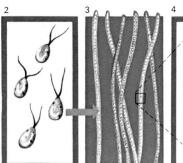 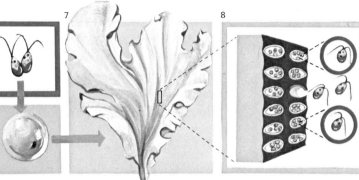 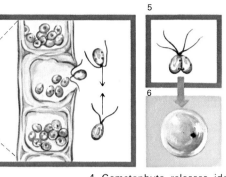

1 Sporophyte is a single cell called a zygote. 2 Sporophyte releases its spores. 3 Each spore grows into a gametophyte plant. 4 Gametophyte releases identical male and female gametes. 5 Gametes swim together and fuse. 6 Fused gametes become zygote, the new sporophyte. 7 The cycle begins again.

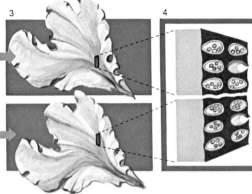 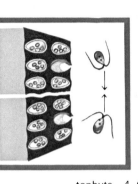 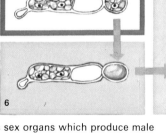 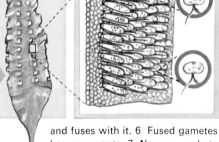

1 Sporophyte plant. 2 Sporophyte releases spores. 3 Each spore grows into large game-tophyte. 4 Gametophytes release identical 'male' and 'female' gametes. 5 Gametes swim together and fuse. 6 Fused gametes become zygote. 7 Zygote has grown into new sporophyte plant. 8 The cycle begins again.

 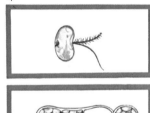 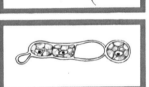 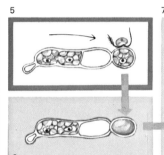

gametes. The female gametes are larger than the male ones.
1 Sporophyte plant. 2 Sporophyte releases asexual spores. 3 Spores grow into tiny male or female gametophyte plants. 4 Male game-tophyte bears sex organs which produce male gametes, female gametophyte bears sex organs which produce female gametes. 5 Female gamete remains attached to its gametophyte. Male gamete swims to female and fuses with it. 6 Fused gametes become zygote. 7 New sporophyte has grown from zygote. 8 The cycle begins again.

 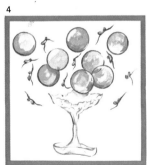 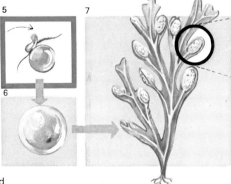 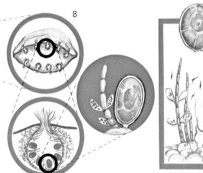

1 Sporophyte has special chambers embedded in fertile regions of its surface. 2 Inside a chamber (a) are many hairs (b) and on some of these are the male and female sex organs (c). 3 Sperm and eggs are released into the water, the females still surrounded by a protective case. 4 Once in the water the eggs are finally freed. 5,6 Sperm swim to the eggs and fuse with them to form zygotes. 7 Zygote has grown into new sporophyte. 8 The cycle begins again.

Plants without colour: the fungi

Another group of plants, the fungi, were evolving in the seas with the algae. Their pattern of life is unlike that of other plants and their structure is also different. They usually grow as a system of tiny, branching threads and are often invisible to the naked eye unless the threads become packed together to form fruit-producing parts such as mushrooms and toadstools. Fungi have no chlorophyll for harnessing the sun's energy, so they cannot produce their own food by photosynthesis. They are therefore completely dependent on outside sources. Some feed on organic materials dissolved in water; others, called saprophytes, break down the tissues of dead plants and animals. Still others depend on living hosts for their food. If they harm the host they are called parasites but if they live in harmony with or even help the host in some way they are called symbionts and the association is called symbiosis.

Fungi play a very important role in plant life for as decomposers they release carbon dioxide back into the atmosphere and return nitrogen to the soil to be reused. Though less conspicuous than other plants and animals, fungi must have had great influence on the evolution of all living things.

The evolution of the fungi themselves is very difficult to unravel. Very few have any hard parts so they are even less likely to be preserved than most plants. However, primitive fungi have been found in the earliest fossil-bearing rocks, the Pre-Cambrian cherts, so we know that they began to evolve in the early seas, alongside the first primitive algae. These first fungi were of the type now called the lower fungi which grow as simple threads, not divided into separate cells. Some still live in water while others have evolved into parasites or saprophytes.

We do not know when the first lower fungi left the sea and became adapted to the different conditions of the land but it was probably at about the same time as the first simple vascular plants and animals, around 400 million years ago. Before that, the land surface was bare and there would have been no easily obtainable organic food for fungi to use. Today they are highly successful on land and the air is constantly full of their small, light spores which germinate easily on a variety of substances, turning food mouldy and causing innumerable infections.

The higher fungi also grow as threads but they have more complex cellular bodies and reproductive structures. They evolved from extinct forms of red algae which lived as parasites on other algae in the ancient ocean. The parasites gradually lost their colour and as they

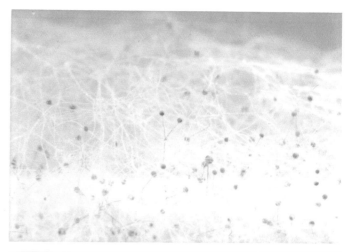

Fly agarics are saprophytes, invading the tissues of dead plants and animals with a tangled mass of threads. Above ground, the threads mass together in a highly organized way to form a spore-producing toadstool.

Top: Bread mould. A dense network of threads forms the body of the fungus. Sporangia are formed at the tip of tiny, upright stalks.

became unable to photosynthesize, they took more food from their hosts and became true parasitic fungi.

The green algae from which the first land plants evolved were probably already infected with these primitive fungi. As plants moved onto the land, the fungi were able to adapt with them, spreading as the vascular plants spread and undergoing a similar evolutionary explosion in the new environment.

The higher fungi can be divided into two groups, the sac fungi, which include the yeasts and the club fungi which include mushrooms and toadstools.

Many of the sac fungi are minute and are only noticeable by the effects they produce on their hosts – powdery mildew, apple scab and Dutch elm disease for example. Damaging green and black moulds are mainly sac fungi but so also are commercially important yeasts and *Pennicillium*, the source of pennicillin. Some sac fungi are saprophytes, breaking down dead logs and leaf moulds: truffles and morels are the fruiting bodies of tiny underground sac fungi.

Puffball releasing a cloud of spores into the wind. The spores will eventually settle and, when conditions are suitable, will grow a new generation of fungal threads.

Lichens are dual organisms. The main body is a fungus but unlike other fungi it does not feed on decaying matter. Within its tissues are chlorophyll-containing cells of a green or blue-green alga. These photosynthesize and provide food for the fungus. In return, the fungus provides the alga with a place to live. Very often lichens are brightly coloured by pigments which mask the green chlorophyll.

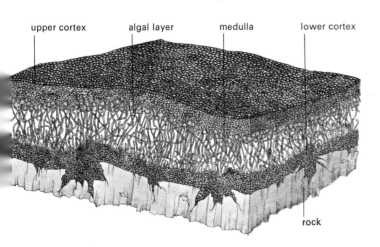

upper cortex algal layer medulla lower cortex

rock

Again, fossils of sac fungi are rare but there is evidence that they were alive in the Carboniferous period. In the Cretaceous there are many more fossils and in the Tertiary even more – possibly because by this time there was a much greater variety of host plants for them to make use of.

The club fungi are the most advanced of all and include both harmful and useful forms. Most are saprophytes. Some, like dry rot fungus, cause serious damage to wood used in buildings – to them it is just dead organic material to be broken down. Others live in association with the roots of trees either as symbionts or parasites. Sometimes the association is so close that the host plant cannot survive without the fungus. Pines, larches and birches all have important fungal partners which have probably improved their chances of survival. Two important plant attackers are also club fungi – the rusts and smuts that cause so much damage to cultivated food crops.

Club fungi probably evolved from some kind of red algae and the earliest forms may have been like rusts. Their hosts, of course, would have been algae instead of vascular plants, for these had not yet evolved. Like the other fungi, they probably began to invade the land with their primitive hosts, the first land plants. Some may already have lived as saprophytes in the water, moving onto land when there was sufficient dead organic matter to support them. Strands of club fungus have been found in fossil wood of a Carboniferous tree fern and many other plants must have been infested by that time – but the fossil evidence is very scarce indeed.

The fungi include the most striking of all examples of symbiosis: the lichens. Lichens are in fact combinations of two plants, a fungus (usually a sac fungus) and an alga. When fungi and vascular plants are symbionts, the fungus is usually the less obvious partner but in the case of the lichens, the fungus is dominant and determines the appearance and form of the dual plant.

Lichen fungi do not feed on dead or decaying organisms: the algal partner is able to photosynthesize and provides enough food for both itself and its fungus. In return, it gains a safe, protected environment.

Lichens are now very widespread but there is little fossil evidence to show how or when they originated. There are, however, signs that they had evolved by the Cretaceous (136-64 million years ago), at about the same time as other sac fungi began to be more numerous. Although they are probably a relatively recent evolutionary development, they are now so well adapted that they are usually the first plants to colonize barren areas and they can survive in the most severe conditions.

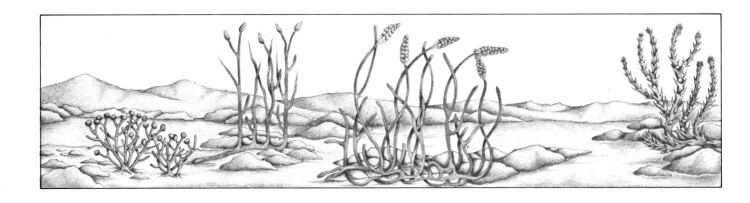

Moving onto the land

Towards the end of the Silurian period, about 410 million years ago, there occurred the most dramatic event in the history of plant life. This was the evolution of land plants.

By this time the early plants had produced, by photosynthesis, an atmosphere around the Earth with 2% oxygen. This, together with its product ozone, prevented ultraviolet rays from reaching the surface. But there were still two main problems to overcome before plants could effectively leave their aquatic habitats: they needed to be self-supporting and they had to be able to withstand the drying effect of the air.

Supported by the water in which they grew, early plants had developed wide branching shapes like large, thin fronds – that is they had a large surface area to volume ratio. Because of the thinness of the plant, gases and minerals could pass freely through the entire surface and diffuse from cell to cell. Little was needed in the way of support. Once on land, however, they had to stand on their own. To do this they needed a more compact shape (a smaller surface area to volume ratio) and a stronger stem. The outer tissues of the stem became more swollen and extra cellulose thickening developed in the cell walls.

To withstand the drying effects of air, a hard outer layer, the epidermal cuticle, developed on the exposed shoot and this prevented too much water from being lost.

As it also prevented gases from entering and leaving the plant, small openings, the stomata, developed to allow and regulate the exchange of gases between the air and the interior of the plant. Water and minerals could now be absorbed only through the rooting system instead of through the whole of the plant's surface, so there had to be a way of transporting these as well as of carrying food manufactured during photosynthesis to the growing parts. The conducting systems that evolved, the xylem for water and minerals, the phloem for food, also helped to support the aerial shoot. Plants with this internal transport system are known as vascular plants.

Not all the plants that attempted to colonize the bare, unoccupied land were successful. One group of branching algae, the Nematophytales, have been described as simple land plants, although they had not developed a xylem and phloem system. Some were quite large, over a metre long and as much in diameter. Some grew in colonies, apparently unattached, probably rolling round in shallow water or on the mud flats. Others were wide and flat, lying on the mud above the water-line. Fossil spores show that at least some of these plants were beginning to make adaptations to life out of water for they had resistant coats which would have enabled them to be dispersed by the wind without drying out.

The Nematophytales failed to adapt fully to the land and were eventually doomed to extinction. What, then,

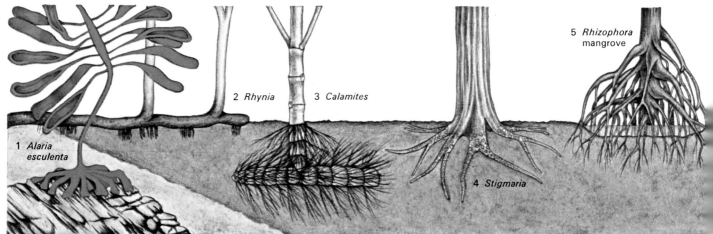

1 *Alaria esculenta*
2 *Rhynia*   3 *Calamites*
4 *Stigmaria*
5 *Rhizophora* mangrove

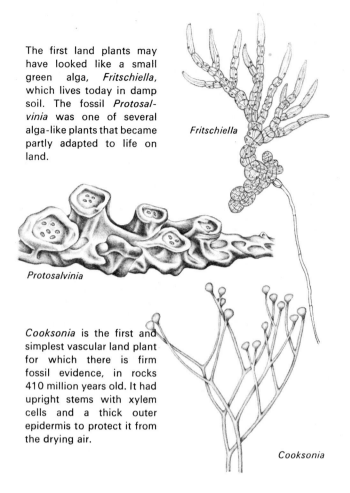

The first land plants may have looked like a small green alga, *Fritschiella*, which lives today in damp soil. The fossil *Protosalvinia* was one of several alga-like plants that became partly adapted to life on land.

Fritschiella

Protosalvinia

Cooksonia is the first and simplest vascular land plant for which there is firm fossil evidence, in rocks 410 million years old. It had upright stems with xylem cells and a thick outer epidermis to protect it from the drying air.

Cooksonia

were the ancestors of modern land plants – and why did they succeed?

In fact there is no fossil evidence for the successful ancestral group though their immediate descendants can be traced so we know that they must have existed and that they were different from the nematophytalean types. They certainly evolved from green algae as today's land plants are still biochemically similar to these; and the fact that they were never preserved probably means that they were small, with no decay-resistant cells. However, the best way we can build up a picture of what they were like is by studying primitive plants that are still alive. A green alga called *Fritschiella* which grows in damp soil is thought to be the closest model. Like *Fritschiella* the new plants probably had two sections, a flat plate of cells fixed to the ground by hair-like rhizoids and an upright, branching 'stem', probably with the reproductive organs. We do not know what colour they were but at least part must have been green, coloured by the pigment chlorophyll which enabled them to trap light for photosynthesis.

The first land plants

The algae from which the first land plants evolved had probably already developed life cycles with gametophyte and sporophyte generations. Most probably the gametophyte and sporophyte would have looked different, the gametophyte lying along the ground, the sporophyte also developing upright branches which carried its asexual reproductive organs. The spores would therefore be released higher above ground level and could be dispersed over a wider area. Gametophyte plants needed to remain small as their free-swimming sperm still depended on a film of water, usually from rain, to reach the egg cells.

The problems associated with leaving the water

Below: All plants need to absorb water and minerals. Algae, which normally live in water, do this through their whole surface though some, like *Alaria* (1) have a root-like holdfast to anchor them. Land plants have developed underground systems for anchoring and to take up moisture and over millions of years these have become adapted to different environments. 2 Underground stem or rhizome for anchoring with rhizoids to take up water. 3 Rhizome with true roots for extra anchoring which in turn bear root-hairs for absorbing water. 4 Underground branches or rhizophores also gave extra support though strictly they were not roots. 5 Stilt roots and (6) breathing roots of coastal mangroves allow root systems to obtain air at low tide. 7 Roots growing down from plant lodged on branch of tree. Eventually these dangling roots strangle the host tree. 8 Aerial roots support heavy branches. 9 Buttress roots act like guy ropes to hold tall trees upright. 10 Food storage root of desert plant. 11 Fine roots spread out to absorb moisture from different levels.

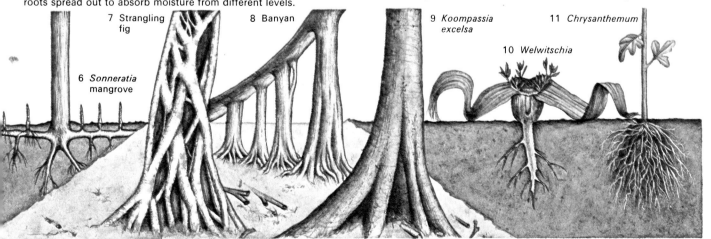

7 Strangling fig   8 Banyan   9 *Koompassia excelsa*   11 *Chrysanthemum*

10 *Welwitschia*

6 *Sonneratia* mangrove

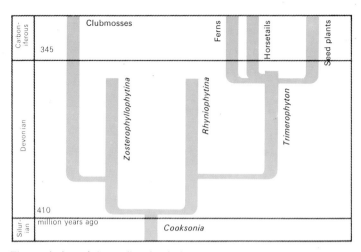

The evolution of the earliest land plants. This family tree shows how the main groups may have been historically related. It shows two major lines of evolution from *Cooksonia*, one leading to the club-mosses, the other leading through *Trimerophytina* to ferns, horsetails and seed plants.

were relatively soon overcome for we know of four places where plants invaded the land at this time. In Bohemia (Czechoslovakia), Podolia (USSR), South Wales and New York State, there were once extensive tidal mud flats which were periodically flooded and dried out as the water level of the shallow Silurian sea fluctuated. Conditions like these were apparently ideal for an evolving land plant because similar plants occurred almost simultaneously in each of these places. These first land plants, called *Cooksonia*, were small with smooth, cylindrical, simply branched stems. All the fossils so far discovered are of sporophyte plants for at the end of the stems are round sacs called sporangia which produced the spores for asexual reproduction.

Cooksonia was clearly the most successful of the land invaders and was the first of an entirely new kind of plants which were no longer algae. It was from such a humble beginning that all the major groups of land plants we know today originated. In the following geological period, the Devonian (410 to 345 million years ago) more and increasingly complicated fossil plants appeared, showing how the plant form is capable of many and varied evolutionary changes. Perhaps in such simple plants slight genetical changes produced more obvious alterations in appearance than they would in today's more complex forms. There was also little competition for the immense areas of land available for colonization and many different forms were able to exist side by side. Recently described fossils from Canada have shown that some of the land plants had many of the primitive anatomical modifications we would expect – stronger stems, conducting strands (the xylem and phloem system), an epidermal cuticle and stomata.

Three major plant groups emerged in the Devonian period: the Zosterophyllophytina, the Rhyniophytina and the later Trimerophytina, together known as the psilophytes.

The Zosterophyllophytina grew at first as clusters of simple, leafless stems about 20cm high. As time passed, more and more of the barren land was occupied and new, increasingly complex forms developed.

As the plants grew taller and thicker, they began to develop simple, spiny leaves which grew out of the stem and provided a larger green surface area for photosynthesis. The spines became gradually larger and flatter until eventually they needed an extension of the stem's xylem and phloem system to carry food, water and minerals to and fro.

The second important group, the Rhyniophytina, were variations on the original *Cooksonia* type of plant.

Some of the land plants that grew during the Devonian period. They show how new plants, more complex than *Cooksonia*, evolved in the two lines that were to lead to modern families.

The sizes given are approximate.

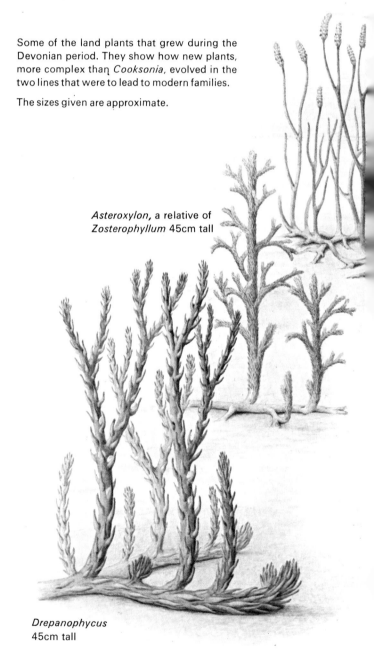

Asteroxylon, a relative of *Zosterophyllum* 45cm tall

Drepanophycus 45cm tall

Next page: About 400 million years ago the area that is now the Brecon Beacons in Wales was part of a newly raised continent. Primitive land plants grew on flat expanses of coastline where rivers and deltas were laying down sediments at the edges of a shallow sea. Some of the plants were better adapted than others for life out of water. 1 *Sporogonites*. 2 *Hicklingia*. 3 *Taeniocradia*. 4 *Drepanophycus*. 5 *Cooksonia*. 6 *Zosterophyllum*. 7 *Sciadophyton*. 8 *Rhynia*. 9 *Gosslingia*. 10 *Dawsonites*.

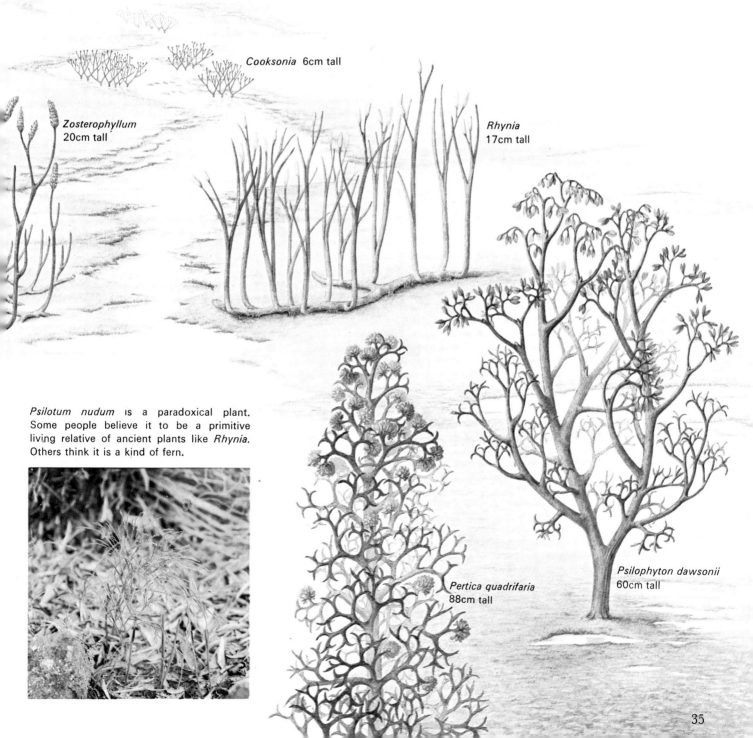

Cooksonia 6cm tall

Zosterophyllum
20cm tall

Rhynia
17cm tall

Psilotum nudum is a paradoxical plant. Some people believe it to be a primitive living relative of ancient plants like *Rhynia*. Others think it is a kind of fern.

Pertica quadrifaria
88cm tall

Psilophyton dawsonii
60cm tall

35

They were rather larger and more branched and their sporangia were egg-shaped instead of round. Within this group is probably the most famous and widely illustrated of all early land plants – *Rhynia*, which comes from the Rhynie chert of Scotland. The cells of these plants are almost perfectly preserved as fossils. Anatomical details are clearly visible, even showing the minute female sex organs (called archegonia) and some strands of fungus which were apparently infesting the plant tissues at the time they were preserved some 400 million years ago.

These early plants reproduced like algae, with two generations. They looked quite different at their different stages. The sporophyte which produced spores consisted of upright shoots rising from a branching stem lying along the ground. The gametophyte had no vertical shoots but was simply a horizontal stem. Both stages were anchored by short hair-like rhizoids; true roots had not yet evolved.

Before the middle of the Devonian, a new group of psilophytes, the Trimerophytina, had evolved from the *Rhynia* group. Like the Zosterophyllophytina they began to form simple leaves, but in a rather different way. Until this time the plants had consisted of stems with shoots growing out in various directions, some regularly, some with no fixed pattern. It is now thought that most leaves evolved on stems where the shoots branched into smaller, unequally sized units which gradually became flattened and were joined together by webbing.

The plants that developed leaves must have gained an advantage over those with only simple stems and in fact these first leaf-bearing trimerophytes were the ancestors of most of the important groups that have existed in the past and that are living today: the ferns, horsetails, gymnosperms such as seed ferns, cycads and conifers and finally, the flowering plants. The trimerophytes themselves, like the rhyniophytes and the zosterophyllophytes, died out.

The first trees

Some Devonian sporophyte plants were becoming taller and larger while they were developing leaves. This increase in size was an attempt either to outcompete others in the struggle for light or to lift their delicate growing points and reproductive organs beyond the reach of ground-dwelling foraging animals such as early insects and amphibians. The gametophyte plants remained small and close to the ground.

Plants could only continue to grow upwards if the amount of supporting tissue in their stems increased.

These two fossils of *Rhacopteris*, a fern from the lower Carboniferous, show how their leaves evolved from branch systems in which the small side branches became flattened and webbed together. In primitive leaves of this kind the original side branches are still represented by the veins. All leaves except those of clubmosses began in this way.

The Australian Devonian fossil *Baragwanathia* is an early clubmoss. Its simple leaves evolved as direct outgrowths of the stem epidermis, each supplied by a single vein.

The larger plants could then branch more – and so needed even more supporting tissue. This interrelationship of increased size, branching and amount of supporting tissue ultimately led to the formation of the woody tree. By the end of the Devonian period, 345 million years ago, all the major groups of plants had evolved larger forms and tree sized clubmosses, horsetails and early gymnosperms dominated the vegetation.

Land plant life cycles

Land plants continued to have sporophyte and gametophyte generations but an important change took place in many of the major groups during the Devonian

Life cycle of the clubmoss *Lycopodium*

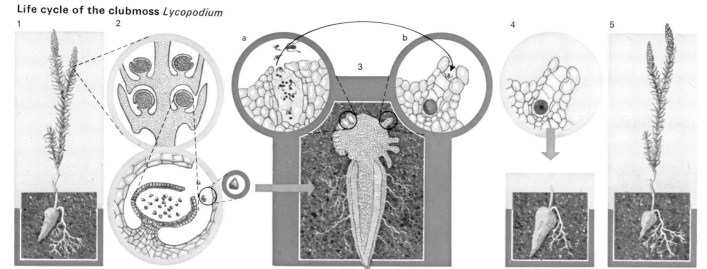

dark blue = haploid   light blue = diploid

Lycopodium is a living clubmoss. Its life cycle shows how the early clubmosses probably reproduced. Like most of the algae they still had sporophyte and gametophyte generations and relied on a film of water to carry the free-swimming male sperm to the female egg cells. *Lycopodium* is homosporous, having spores all of one size.

1 Sporophyte plant. 2 Sporophyte releases spores from special capsules, the sporangia. 3 Each spore grows into small gametophyte plant which develops male (a) and female (b) sex organs. Male organs produce sperm which are released and swim freely. Female organs each produce an egg which remains within the sex organ. Sperm swim to the egg and fertilize it. 4 The fertilized egg stays inside the gametophyte and begins to grow into a new sporophyte plant, first sending out a root and then a shoot. 5 As the new sporophyte grows, the gametophyte withers away and the cycle begins again.

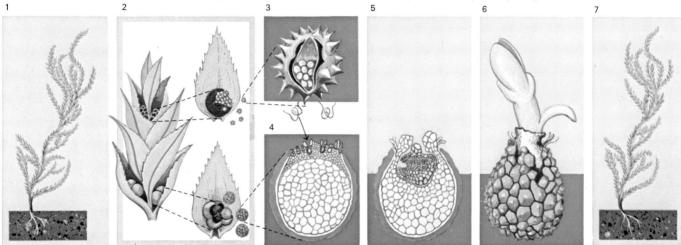

Life cycle of the clubmoss *Selaginella*

dark blue = haploid   light blue = diploid

Selaginella, another living clubmoss, is heterosporous, producing both 'male' and 'female' spores of different sizes. The gametophyte stages are similar to those of *Lycopodium* but are more advanced. Both male and female gametophytes develop within their spore cases and do not grow into independent plants.

1 Sporophyte plant. 2 Sporophyte cone releases small 'male' and larger 'female' spores from different parts of plant. 3 Male spore contains tiny male gametophyte which releases swimming sperm. 4 Female spore, containing female gametophyte, ruptures and the gametophyte protrudes to expose the female sex organs. Sperm can then swim in to fertilize the eggs. 5 The fertilized egg grows inside the female gametophyte, still within the original spore case. 6 The new sporophyte grows out of the spore case. 7 The cycle begins again.

period. Until this period, each sporophyte plant had produced spores of only one kind; they were what is known as homosporous. Now, some plants began to produce two kinds of spores from two kinds of sporangia, an evolutionary stage which some algae were also to reach in the sea. These plants are said to be hetero-sporous.

In heterosporous plants the two different spores germinate into two small but distinctly different gametophytes which remain within the spores and never grow into a large green plant. The two kinds of spores are rather different in size, for a very good reason. The larger type becomes a female gametophyte with female sex organs. It therefore contains the food reserves necessary for the early growth of the future plant. The smaller spores which become the male gametophytes have no food reserves as their sole function is to produce and liberate the sperm before dying.

For heterosporous plants, the sexual part of the life cycle becomes less obvious. The sporophyte plant is increasingly dominant and because the reproductive cycle is speeded up, the next generation can be estab-lished more rapidly. This means that evolution itself can be accelerated as changes can be passed on more quickly and the possibility of new life forms appearing becomes greater.

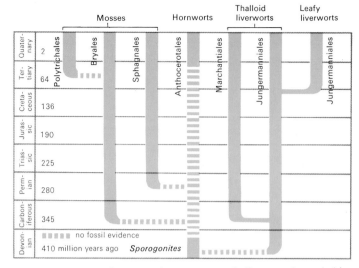

The bryophytes (mosses, hornworts and liverworts) probably originated in the Silurian, at about the same time as the vascular plants. The oldest fossil bryophyte was probably *Sporogonites* but it is difficult to work out details of bryophyte evolution, partly because they were rarely preserved as fossils.

Hornworts, mosses and liverworts

Dry land was colonized by the ancestral vascular plants during the closing part of the Silurian period over 400 million years ago but they were not the only plants to succeed in living out of the water. Another group, the bryophytes, managed to do this during the early part of the Devonian and has continued to evolve and diversify ever since. Both bryophytes and vascular plants were descendants of the green algae so their life cycles are very similar. They look, however, very different and this is a direct result of the different ways in which they invaded the land.

Algae can be either sexually reproducing gametophyte plants or asexual, spore-producing sporophyte plants. As we have seen, their life histories usually show a series of changes leading from one type of plant to the other. Vascular plants evolved as some green algal sporophytes became successful at living on dry land; bryophytes evolved as the gametophytes of other green algae became equally successful.

The bryophytes, then, are gametophyte plants and have to deal with all the problems of sexual reproduction. There is no way that they can grow very large, let alone bush or tree sized, for the sperm are free swimming and need at least a surface film of water to enable them to reach the egg cell, which remains attached to the parent plant. Only quite small plants can retain this film of water.

The earliest bryophytes come from the Devonian and are called *Sporogonites*. The gametophytes grew into flat plate-like plants anchored to the ground by rhizoids. The sperm swam to the archegonia and from the fertilized egg cell grew a small sporophyte which released spores. These spores eventually grew into the next gametophyte plant. The sporophytes consisted of only a spore producing capsule at the end of a long stalk and, unlike the ancestors of the vascular plants, never became independent. Instead they remained as parasites on their parent gametophytes, relying on them for food and water throughout their lives.

This is still the pattern of life in the three living groups of bryophytes: the hornworts, liverworts and mosses. All have small but obvious gametophytes which produce even smaller, dependent sporophytes.

The hornworts are the simplest of the bryophytes and look rather like the earliest fossils, *Sporogonites*. However, there are unfortunately no fossils that link the two together. Liverworts are found all over the world and are very varied in shape, ranging from plate-like 'thalloids' to leafy plants rather like mosses. The thalloid liverworts are very obvious plants, reaching

several centimetres in length in the humid tropics; they grow along the ground or sometimes on trees. Leafy liverworts usually grow along the ground on other plants or among other liverworts and mosses. At first sight the thalloid and leafy liverworts look as different from one another as they are from mosses but there are enough similarities to justify grouping them together.

Mosses are a larger group today than the other two and are the most widespread. They look more like vascular plants than the other bryophytes and have stems and a type of leaf. At the young gametophyte stage they are very like the algae which produced the vascular plants, with some parts running along the ground and others standing upright.

Mosses grow in most places, even in man-made habitats such as on walls, roofs and in paving cracks. Some are even able to survive the high levels of pollution in our cities. Under certain conditions they may become

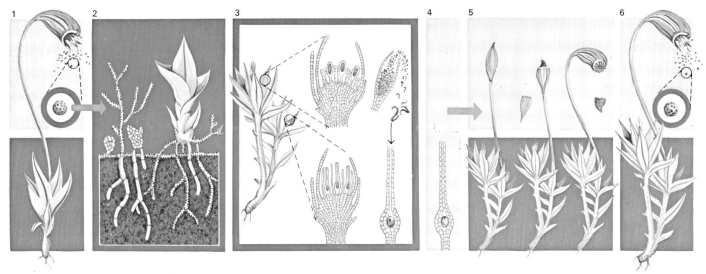

Life cycle of the moss *Funaria*

dark blue = haploid   light blue = diploid

Mosses have spore and gamete producing generations but only the gametophyte grows into an independent free-living plant.
1 Sporophyte grows from gametophyte and releases spores. 2 Spores grow into new gametophyte plant. 3 Gametophyte develops male and female sex organs. Sperm swim in surface water to female sex organs and fertilize eggs. 4 Fertilized egg stays inside gametophyte and begins to grow. 5 New sporophyte plant grows, still dependent on gametophyte. When mature, it develops a capsule which contains the new spores and is covered by a protective cap, the calyptra. This falls away to reveal the operculum. Just before the spores are released, the operculum also falls off. 6 Finally, the ripe spores are released and the cycle begins again.

the dominant plants – the vast bogs of the northern hemisphere are built up of accumulations of dead plants, one of the principal ones being *Sphagnum* moss.

Sphagnum is specially adapted for growth in wet places and has dead hollow cells in its leaves which enable it to store large amounts of water. It thrives in acidic conditions where other plants die and is able actually to make its environment more acid by absorbing certain chemicals selectively from the water. Providing that the rainfall level is high enough, *Sphagnum* will grow and accumulate, controlling its own ideal environment. A high level of acidity also prevents fungi and bacteria from growing so the dead mosses decay very little, building up instead into thick layers of peat.

Although there are no early fossils of the hornworts, there is evidence that the early bryophytes diversified quickly and by the Carboniferous (345 to 280 million years ago) there were recognizable mosses and thalloid liverworts. Leafy liverworts appeared much later, around 64 million years ago. Fossil bryophytes are in fact rare for such small plants had much less chance of becoming fossilized than did the larger and more rigid vascular plants. Occasionally, however, they were well preserved: one of the best collections comes from Permian rocks in the USSR. Some of these fossils are so detailed that they can be compared closely with living families.

These early fossil bryophytes give no clear picture of the evolutionary history of the group but the fact that both mosses and liverworts had appeared by then and were already so different from each other suggests one important fact: the changes that led to their evolution probably happened at least twice and although their life cycles are so similar, they are probably descended from two different groups of green algae which, independently, solved the problem of living on the land in the same way.

Left: The gametophyte stage of a complex thalloid liverwort *Marchantia polymorpha*. The umbrella-like heads are special extensions of the gametophytes, carrying either male or female sex organs. Eventually the sporophytes grow downwards from the female umbrellas before shedding their spores.

Right: Sporophyte stage of the moss *Funaria*, growing from the parent gametophyte.

Far right: Moss carpeting a woodland floor. Mosses never grow taller than 60cm and few grow taller than 10cm, but they can spread rapidly and may become the dominant ground covering plants.

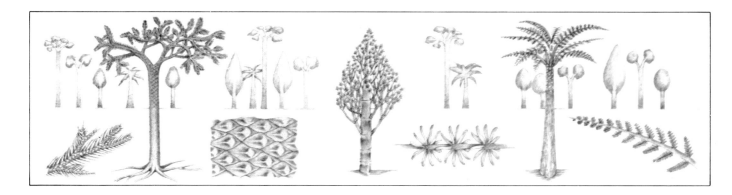

The great coal forests

During the last part of the Devonian period and the following Carboniferous land plants of many different species took over large areas of the land. It was during this time that two major plant groups – the clubmosses and the horsetails – reached the peak of their development. They were descended from different psilophyte ancestors but both went through a similar series of changes, producing more variations upon their basic structures and gradually increasing in size. The name Carboniferous means literally 'coal-bearing' for it was during this period that the great coal forests flourished in many parts of the world.

In the Carboniferous period conditions were ideal for the evolution of many forms of large plants. At this time the continents we know today formed a single landmass. The area of land which is now northern Europe and north-eastern America was a gradually sinking shallow basin of fresh water and was situated much nearer the Equator than it is today. Into this habitat came the clubmosses and horsetails, both ideally suited for such a watery environment. The equatorial climate, which was probably warm and damp, made plants grow luxuriantly. There were few very large land animals and the number of plants was limited only by competition for rooting space, nutrients and sunlight.

With so many plants growing together, the amount of plant debris falling into the water eventually exceeded

the natural rate of decay with the result that the lake gradually turned into a peaty swamp. As the area continued to subside many metres of undecayed or partially decayed plants accumulated. This was the first stage in the formation of coal.

The rate at which the land was sinking controlled the swamp vegetation. When subsidence was slow, the debris often piled up above the general water level, enabling more plants to take root. Here could be found the smaller plants which could not survive in areas covered with water. In contrast, if the subsidence was more rapid, plant growth could not keep pace and vegetation only survived around the edges of the basin. At the same time the central parts of the basin silted up

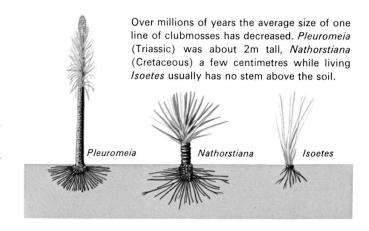

Over millions of years the average size of one line of clubmosses has decreased. *Pleuromeia* (Triassic) was about 2m tall, *Nathorstiana* (Cretaceous) a few centimetres while living *Isoetes* usually has no stem above the soil.

Pleuromeia   *Nathorstiana*   *Isoetes*

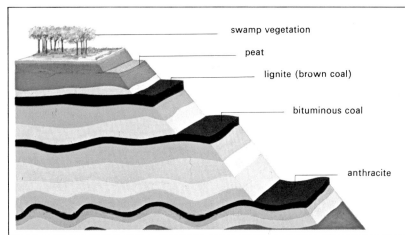

swamp vegetation

peat

lignite (brown coal)

bituminous coal

anthracite

How coal is formed

The dense swamp forests of the Carboniferous with their pools of shallow, stagnant water, were ideal peat-forming areas. Peat, the first stage in coal formation, was a dense layer of plant debris which had accumulated and had not decayed very much. As more sediments were laid down, the peaty debris became compressed into coal and water and gas were gradually forced out. Lignite or brown coal still contains 43% water while bituminous coal has only 3%. Anthracite, the most compressed, has hardly any water content at all.

Period	(million years ago)								
Quaternary	2	*Isoetes*	*Stylites*	*Phylloglossum*		*Lycopodium*	*Selaginella*		
Tertiary	64								
Cretaceous	136		*Nathorstiana*						
Jurassic	190								
Triassic	225		*Pleuromeia*						
Permian	280				*Lepidodendron*				
Carboniferous	345								
Devonian	410 million years ago		*Zosterophyllum*					▭▭▭ no fossil evidence	

Clubmosses and their relatives formed one of the major groups of land plants to evolve in the Silurian-Devonian. By the Carboniferous some had changed from the earlier creeping forms like *Drepanophycus* to large, branching trees like *Lepidodendron,* up to 45m high. Other herbaceous plants also evolved and though the large trees have all long since disappeared, some of the smaller plants have survived to the present day.

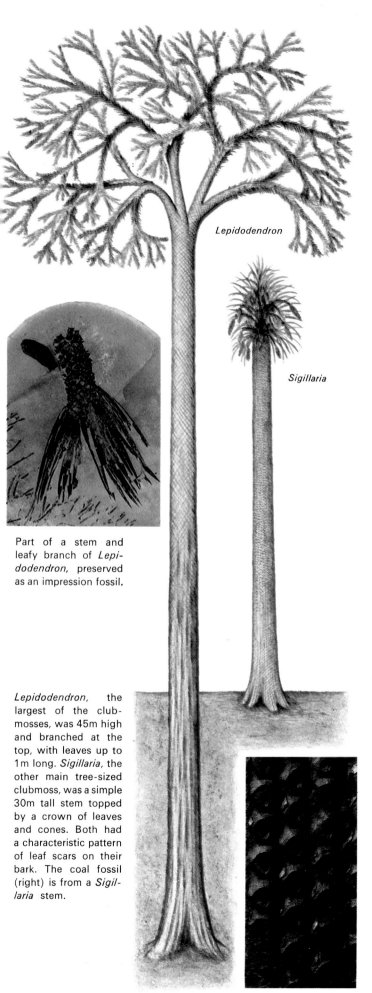

Lepidodendron

Sigillaria

Part of a stem and leafy branch of *Lepidodendron,* preserved as an impression fossil.

Lepidodendron, the largest of the club-mosses, was 45m high and branched at the top, with leaves up to 1m long. *Sigillaria*, the other main tree-sized clubmoss, was a simple 30m tall stem topped by a crown of leaves and cones. Both had a characteristic pattern of leaf scars on their bark. The coal fossil (right) is from a *Sigillaria* stem.

The underground rhizophores of several giant lepidodendrons (left) which were once part of a dense swamp forest. Right: *Lycopodium selago*, a living clubmoss that has probably changed little since the Carboniferous.

again as water and sediments rushed in from the surrounding mountains. As rates of subsidence and silting varied, successive invasions of plants crept out from the edges to cover the entire area – and were successively covered. Below the surface, in the middle of the basin, a series of layers of plant debris built up, separated by sands and silts. As more and more material accumulated on top the layers were compressed and, over millions of years, they hardened into coal.

In such ideal growth conditions, the plants naturally increased in size and adapted themselves even more for existence in the area. The tree sized clubmosses all had tall, specially thickened trunks with only a central woody core and varied from simple stems like *Sigillaria* to types such as *Lepidodendron* with its many branches. They had shallow rooting systems, some called *Stigmaria*, which spread out to anchor them in the mud. Because these clubmosses were the first major colonists of the lakes, they were rooted in the original sediments

Period	Million years ago				
Quaternary	2	*Equisetum*			▌▌▌▌▌ no fossil evidence
Tertiary	64				
Cretaceous	136				*Neocalamites*
Jurassic	190	*Phyllotheca*			
Triassic	225		*Schizoneura*		
Permian	280	*Sphenophyllum*	*Archaeo-calamites*	*Calamites*	
Carboniferous	345				
Devonian	410 million years ago		*Ibyka*	*Pseudosporochnus*	
		Hyenia		*Trimerophyton*	

Like the clubmosses, horsetails produced large trees, *Calamites*, in the Carboniferous period, as well as the smaller *Sphenophyllum* and *Equisetum*. *Phyllotheca* and *Schizoneura* were southern hemisphere horsetails. Species of the single genus *Equisetum* are the only modern survivors of this once large group.

of the lake bottom. Often the sediments were pressed down over the *Stigmaria* before they had decayed and many cast fossils have been found, replicas of the original living organs. The most famous are in the fossil grove in Glasgow where eleven stumps are preserved in their actual growing positions. Such examples show how close together the giant clubmosses grew and how dense the forests must have been.

Tall clubmosses reproduced by releasing large quantities of spores from cones formed high up on the mature plants. Some cones were up to 50cms long and probably produced as many as eight thousand million male spores. The larger female spores were far fewer in number, counted in hundreds rather than millions. Released from heights of up to 45m, the spores could be carried some distance to colonize suitable new habitats, although the wastage rate must have been enormous. We do not know how long each sporophyte plant lived and how many generations of spores it produced but it seems likely that in these ideal conditions they would have lived for many years.

The larger horsetails, such as *Calamites*, were equally successful at rapid colonization although they did not rely entirely on spore production. Like the living horsetail *Equisetum*, they grew underground stems which spread horizontally and produced large numbers of stems up to 10m tall. The stems had regular rings of branches and formed dense thickets which were probably as impenetrable as modern bamboo thickets.

Both clubmosses and horsetails experimented with their methods of reproduction to achieve faster and better seedling growth. Larger spores evolved containing greater food stores which enabled the new plant to grow protected from the pressures of the environment. Eventually the spores became so large that there was only room for one in each spore case. They were now

Annularia, the fossil leaves of the giant Carboniferous horsetail *Calamites*, which grew up to 16m high.

Calamostachys, a calamite cone.

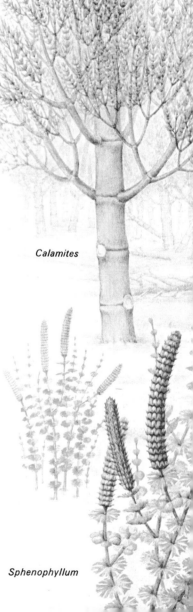

Calamites

Sphenophyllum

44

in many ways similar to the seeds of living conifers and flowering plants but we shall see later that these groups have much more complex reproductive organs.

The adaptations of size and reproduction that enabled these giant plants to dominate the swamps also made them completely dependent on one particular environment for their survival. When the swamps dried up and disappeared at the end of the Carboniferous most of the giants were too specialized to survive and became extinct. These were the first of many extinctions brought about by a changing environment – extinctions mirrored millions of years later in the animal world by the rapid disappearance of the dinosaurs and larger carnivorous mammals.

Next page: A coal swamp in Europe, about 300 million years ago. Sometimes the swamps were covered with dense forests of giant club-mosses (*Lepidodendron* 1) but at other times they were more open with areas of deep, stagnant water. Other clubmosses were *Sigillaria* (2) and *Lepidophloios* (3). Giant horsetails (*Calamites* 4) tree ferns (*Psaronius* 5), seed ferns (*Medullosa* 6), and herbaceous plants such as ferns (*Zeilleria* 7), horsetails (*Sphenophyllum* 8) and small clubmosses (*Sellaginellites* 9) may have grown in the swamps and also in nearby floodplains and deltas.

A few clubmosses and a single horsetail, *Equisetum*, are still alive today, the only survivors of much larger families. *Equisetum* and the clubmosses *Lycopodium* and *Selaginella* are virtually the same as plants that grew among the now extinct giants. The small clubmoss *Isoetes* is a dwarf descendant of tree sized ancestors: most of the plant is below the soil with only its long grass-like leaves visible above ground.

Smaller plants existed alongside the giants of the Carboniferous swamps although they grew mainly in the drier, more raised areas. Here were the ferns, varying in height from small plants to tall, graceful tree ferns such as *Psaronius*, an 18m high giant with leaves up to 3m in length. Like the clubmosses, many of these ferns are very close to living species though their modern relatives do not have such large stems.

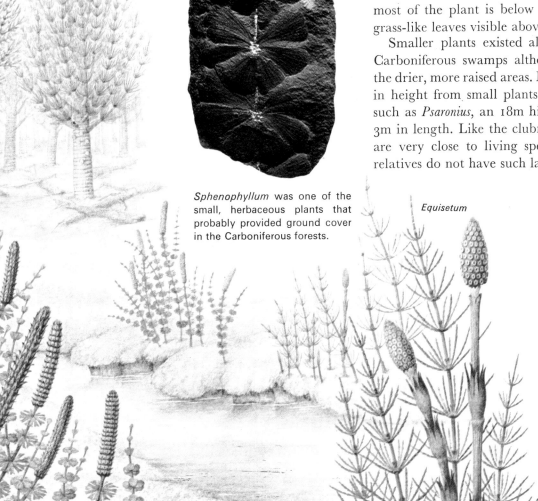

Sphenophyllum was one of the small, herbaceous plants that probably provided ground cover in the Carboniferous forests.

Equisetum

Fossil stems of *Equisetum*, buried quickly by sandy sediments in the middle of the Jurassic period and preserved in their growing position.

The first ferns had evolved from the Trimerophytina group of psilophytes in the Devonian. In the early Carboniferous some were still passing through the first stages of leaf evolution and had three dimensionally branched leaf systems which are very difficult to distinguish from shoots. Ferns such as *Cladoxylon* and *Stauropteris* survived until late in the Carboniferous period. Most Carboniferous ferns, however, were much more advanced than their Devonian ancestors and at first sight it is difficult to realize that they are related. They already looked much more like modern ferns, with two dimensional webbed leaves. For a time they lived alongside the more primitive forms but eventually these disappeared, outcompeted by their more adaptable relatives.

Ferns have been found in all subsequent geological periods and a study of their fossils shows that the group is a continually evolving and successful one. The picture that has been built up shows a repeated series of changes where new groups of ferns expand in both numbers and geographical distribution, only to go into decline later on. Different groups also reached their maximum size at different periods.

Carboniferous tree ferns such as *Psaronius* went into a decline after the next period, the Permian (280 to 225 million years ago) and there are now only about two hundred species of this kind, growing in the tropics. The royal ferns which appeared in the Permian were at first quite small plants: they produced their largest members during the Cretaceous (136 to 64 million years ago) when the South African *Osmundites kolbei* probably had a stem 1m tall. After this their average height decreased until today their stems are only a few centimetres high. Interestingly, though, a close relative of the royal ferns, *Leptopteris wilkensia*, grows to a height of about 3m, perhaps representing a later stage of enlargement.

Most living tree ferns are not the direct descendants of the Carboniferous giants but belong to two separate families which evolved in the middle of the Jurassic period, some 163 million years ago. They are found in all tropical lands south to Chile, New Zealand and South Africa and, up to 10m high, are probably now at their 'large' stage. The most recent group to show a marked size increase is *Blechnum*. It has not yet reached its giant phase but one of its species, *Blechnum gibbum*, has a stem nearly 50cm tall and may be the first stage in the development of a tree fern of the future.

All mature tree ferns taper upwards to end in a collection of large leaves but they achieve this shape by a rather curious pattern of growth. The growing tip of the trunk gradually increases in size as the plant gets

Right: *Psaronius* was one of the largest of the Carboniferous tree ferns, about 18m high with leaves up to 3m long. The leaves grew at the crown, dying away as the stem lengthened but leaving their tough bases as a kind of bark-like covering.

Angiopteris, one of the largest and most primitive living ferns, may be a descendant of tree ferns such as *Psaronius*. It no longer has a tree-like stem but its fronds may reach 7m in length.

Stauropteris

Cladoxylon and *Stauropteris* were among the first ferns to evolve in the Devonian and Carboniferous periods. *Cladoxylon* probably grew to about 50cm high; *Stauropteris* was larger and more bush-like, about 1m high. Both had primitive leaves which were plainly modified branch systems and produced spores from sporangia at their tips. *Stauropteris* was heterosporous and shows how the first seeds may have started to evolve.

Cladoxylon

Psaronius

Tempskya

Tempskya was a large tree fern, some 18m high, which grew in the Cretaceous period over 64 million years ago. Its upper part was covered with a dense mass of small leaves. The trunk of this strange plant consisted of a complex mass of narrow roots and wider stems, intertwined together.

Below: Part of a cross section of a fossilized Tempskya trunk showing the many narrow roots and relatively few, wider stems, preserved in microscopic detail.

Blechnum gibbum is a more recently evolved tree fern and has yet to reach its giant stage.

Osmunda kolbei

Royal ferns first appeared in the Permian, over 250 million years ago. Osmundites kolbei grew in South Africa in the Cretaceous and was much larger than today's royal ferns, with a stem up to 1m tall. Below: Fertile fronds of living Osmunda regalis, the royal fern.

Asplenium creticum is one of the youngest of all ferns, a small, creeping plant found today only on the island of Crete.

Blechnum gibbum

Asplenium creticum

49

Life cycle of a fern

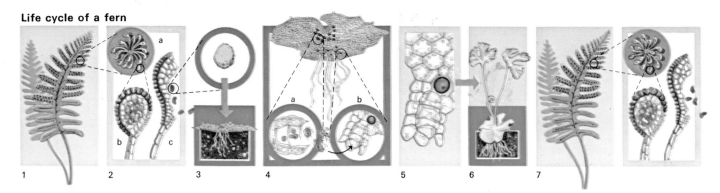

dark blue = haploid   light blue = diploid

The leafy plants we call ferns are their sporophytes. The gametophytes which produce sperm and eggs remain relatively small for the free swimming sperm still rely on surface water from their damp surroundings to carry them towards the egg cells.

1 Sporophyte plant bears spore-producing sporangia, usually on the backs of the leaves. 2 Spores (a) develop inside sporangia (b) which split open (c) to release them. 3 Each spore grows into a small gametophyte. 4 The mature gametophyte develops male (a) and female (b) sex organs on its underside. Sperm swim to egg inside female organ and fertilize it. 5 Fertilized egg begins growth inside gametophyte. 6 New sporophyte plant develops on top of the gametophyte and gametophyte withers. 7 The cycle begins again.

taller. This would make the tree an inverted cone and to prevent this large numbers of roots grow out of the lower trunk. The dense mass of roots around the stem makes it very wide at the base and provides its main means of support. *Tempskya*, a tree fern from the Mesozoic era (225 to 64 million years ago) was even more unusual for the lower part of its 6m tall trunk seems to have consisted entirely of roots. The central stem apparently died and decayed early in the plant's life so that it left no recognizable traces. The supporting roots growing from higher up fed water and minerals directly into the upper trunk. *Tempskya* was an important plant in the northern hemisphere during the Cretaceous period but then disappeared, possibly affected by changes in climate. No close ancestral forms exist today.

Evolutionary adaptations

Most ferns have always been much smaller than the tree ferns. There are about 10,000 species alive today, ranging widely in size, shape and geographical distribution. There are delicate textured, filmy ferns which can only grow in areas with high humidity and others – like the rusty-backed fern – which can survive long periods of drought and can even grow on dry stone walls. Some are lime lovers, others avoid limy soil and only grow in acidic areas. A few can only be found within reach of sea air. Some species live only in warm, tropical climates, others in cool, temperate regions or in cold mountain habitats. A small number are epiphytic, that is they grow on other plants, and some live actually in water. All these specializations are evolutionary adaptations which have enabled different plants to colonize new environments and avoid competition with other species.

Of the vascular plants the adaptable ferns are surpassed in number and variety only by the flowering plants. Their success depends on several factors, one of which is their method of reproduction.

Ferns are still today colonizers of new habitats. Above: Clover-like *Marsilea* roots in muddy marshes and shallow pools. Below: A small fern (*Polypodium pellucidum*) and a flowering plant (*Coprosma ernodeoides*) have found footholds in a lava flow on Hawaii Island.

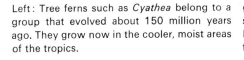

Left: Tree ferns such as *Cyathea* belong to a group that evolved about 150 million years ago. They grow now in the cooler, moist areas of the tropics.

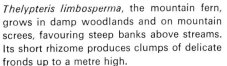

Thelypteris limbosperma, the mountain fern, grows in damp woodlands and on mountain screes, favouring steep banks above streams. Its short rhizome produces clumps of delicate fronds up to a metre high.

Most land ferns produce large numbers of spores which are small and light enough to be carried over vast areas by the wind. Unlike many plants, different fern species are able to cross or hybridize naturally. This may produce more successful offspring, able to compete more favourably in the habitat or to move into new areas. Many species can spread without having to go through the normal reproductive cycle: they may grow underground stems like bracken or, like the sword ferns, produce surface runners similar to strawberries. Others form small buds called bulbils which can take root and grow into separate plants. Some such as *Asplenium viviparum*, produce bulbils all over the leaves while others such as the walking fern *Camptosorus*, form them at the leaf tips.

The water ferns are a small but important group consisting of five genera. Some of them, such as *Marsilea*, root in wet or marshy places while others, such as *Salvinia*, float freely on the surface of the water. All other living ferns produce spores of only one type but water ferns are heterosporous. Both male and female sporangia

are embedded in hard-walled, bean shaped structures called sporocarps. Because these ferns can only grow in watery habitats, long dry seasons are very dangerous for them and the sporocarps are adaptations which enable them to survive. Inside their protective wall, the sporangia can lie dormant in dry conditions for up to thirty years. When the habitat becomes wet again and new plants have a better chance of establishing themselves, the sporangia become active.

This evolutionary adaptation for life in wet places is probably relatively recent for the earliest fossils come from the Cretaceous. As only fossil spores have been found it is impossible at the moment to be sure what their ancestors were or to work out how they are related to the more common homosporous ferns.

Although the ferns have never dominated large areas or complete habitats in the way clubmosses and horsetails dominated the Carboniferous swamps, they are found in nearly every possible community and are often important parts of the vegetation. They are clearly plants which are still changing, evolving and expanding.

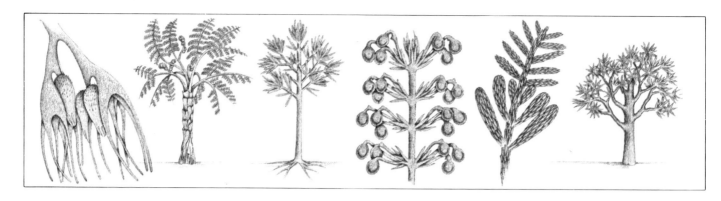

The first seeds

During Devonian times, as the clubmosses, horsetails and ferns were emerging, another major group was evolving and becoming important in many areas. These were the immediate ancestors of the gymnosperms. The word gymnosperm means literally 'naked seed' and the group includes conifers, cycads, maidenhair trees and a few other less well known plants. Their ancestors were the progymnosperms.

We know very little about the beginnings of the progymnosperms, but evidence seems to show that they all evolved from a common psilophyte ancestor – from one of the early land plants that had lived in the Devonian period. The oldest and simplest is about 370 million years old. Named *Tetraxylopteris*, it was a shrubby plant with a three-dimensional branching system and early forms of unwebbed leaves. The fertile branches, on which grew clusters of sporangia, were also arranged in a complex three-dimensional pattern.

There are many other genera and species of progymnosperms in rocks from the later Devonian and early Carboniferous periods and they may even have survived into the later Carboniferous before becoming extinct. These strange plants had woody stems like gymnosperms but the later forms had leaves which looked more like ferns. Indeed, it was not until 1960 that leaves and stems were proved to belong to the same plant and the group could be recognized as a distinct entity.

The first progymnosperm to be reconstructed was a large plant known as *Archaeopteris*. It was clearly a tree, probably looking rather like a 30m tall Christmas tree. The large branches were divided into many leaves and some of these carried the sporangia. Some species had sporangia of two kinds, proving that some of the progymnosperms were heterosporous. This is important for the production of two kinds of spores is a vital stage in the evolution of seeds. From these simple beginnings came the gymnosperms which, in their turn, were the ancestors of the flowering plants.

The evolution of seeds

Scientists now believe that a series of changes brought about the evolutionary advance from heterosporous spore formation to seed production, where the gametophyte, though still essential, is protected by the sporophyte and is almost invisible to the naked eye. The familiar trees, shrubs and flowering plants we see around today are all seed plants.

To produce seeds each 'female' sporangium makes one large 'female' spore instead of several and retains it permanently on the parent plant. The sporangium (now called the nucellus) grows an extra protective layer, the integument, which leaves only the tip exposed to receive the male spore. The modified sporangium is an ovule, the correct botanical name for an unfertilized seed. Inside the ovule a tiny female gametophyte forms.

The smaller 'male' spores produced by heterosporous plants now become pollen grains. They are released by the male sporangia (the pollen sacs) and carried away by the wind. These pollen grains are all that remain of the male gametophyte plant. Some reach ovules and there become trapped in watery droplets called pollen drops which form at the tip of the ovules. The ovule reabsorbs its pollen drop and as it does so, carries the pollen grain inside its protective integument. Once

Life cycle of the cycad *Dioon*

dark blue = haploid   light blue = diploid

Cycads are living, primitive seed plants with a method of fertilization slightly more advanced than that of their extinct forerunners. Pollen grains anchor themselves to the ovule opening and grow towards the egg cells. In the final stages sperm swim to the egg but for such a

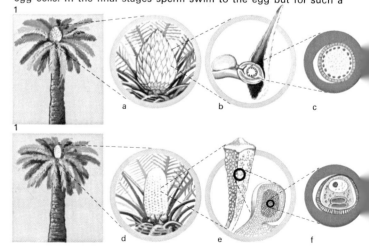

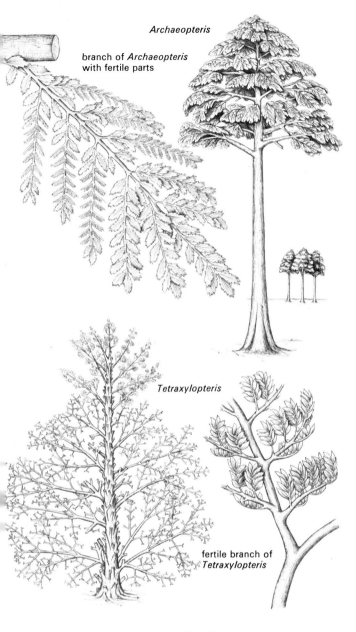

Archaeopteris

branch of *Archaeopteris* with fertile parts

Tetraxylopteris

fertile branch of *Tetraxylopteris*

The progymnosperms were Devonian relatives of the first seed-bearing plants. *Archaeopteris* was a woody, branched tree, some species of which grew up to 18m tall. The branches bore frond-like twigs with spirals of green leaves (far left) The fossilized frond (below) is from a tree which grew in Ireland during the Upper Devonian period, about 350 million years ago. Some species were heterosporous, producing two kinds of spores — a vital step in the evolution of the seed. It is possible that the homosporous species were really the male trees and that the females bore the seeds.

Tetraxylopteris was a smaller, shrubby plant, probably about 2m tall. Its spores were formed on fertile branches (left) which grew on the upper part of the bush.

short distance that the chances of fertilization are good.

1 Male and female cones grow on separate trees. Female cones (a) have scales (b) each with 2 ovules (c). Males cones (d) have sporangia (e) which release pollen (f). 2 Female cone (a) opens to receive pollen. Inside the cone (b) the ovule, containing the female gametophyte, traps pollen in its pollen drop. 3 Six months later the cone is tightly closed (a). Inside the ovule (b) the pollen has grown a tube towards the egg and released sperm. The sperm swim to the egg and fertilize it. 4 The cone remains closed (a) while inside the ovule (b) the embryo grows. 5 Some months later the cone scales open (a) to release the ripe seeds (b). 6 The new seedling germinates.

7 When the tree is mature, the cycle begins again.

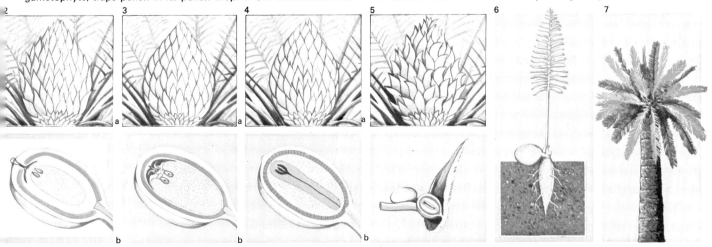

inside, the pollen grains liberate their swimming male sperm into the remains of the water. Meanwhile the tip of the nucellus breaks down to expose the female gametophyte tissue, called the prothallus. Here the archegonia attract sperm to the egg cell in the usual way. Fertilization takes place inside an archegonium and an embryo forms there. The ovule now becomes the seed, carrying the embryo plant of the next generation, protected by the dried and hardened integument.

The hardening of the seed wall effectively isolates the developing embryo within its food store and it is then shed from the parent plant. If conditions are suitable the seed will germinate and the young seedling will become established and begin to grow. Primitive seeds must germinate quickly if the young plant is to survive but millions of years of evolution have produced plants with seeds that may remain dormant for long periods of time. They can therefore overwinter or survive periods of drought, delaying germination until the seedling has more chance of growing successfully.

A new root is the first sign of growth for the plant needs to be securely enclosed in soil where it can obtain its necessary supply of water. It uses the food reserves of the seed for this early growth; only later, when its leaves appear, does it begin to photosynthesize and produce its own food.

This type of life cycle has many advantages over those that include a separate, free growing gametophyte stage. The environment has less influence over the growth of the gametophyte and the sperm is no longer dependent on external water to reach the egg cell. After fertilization the embryo can grow and wait safely inside the seed until conditions are right for germination. Enclosed in its protective coat it can survive dispersal by wind, water or animals, drought or cold periods more easily. In the earlier life cycles the embryo had to grow soon after fertilization, regardless of any changing environmental conditions. In the case of the large clubmosses and horsetails this proved disastrous.

The first seeds

It is now thought that seeds evolved in several different types of progymnosperms, leading to several independent groups of gymnosperm plants. Some of these, we shall see, were much more successful than others.

The oldest seed-containing fossils known so far come from late Devonian rocks in Pennsylvania, USA. They are called *Archaeosperma* but unfortunately we do not know which plants produced them. Fragments of the progymnosperm plant *Archaeopteris* were found with them but we do not know for certain that this was

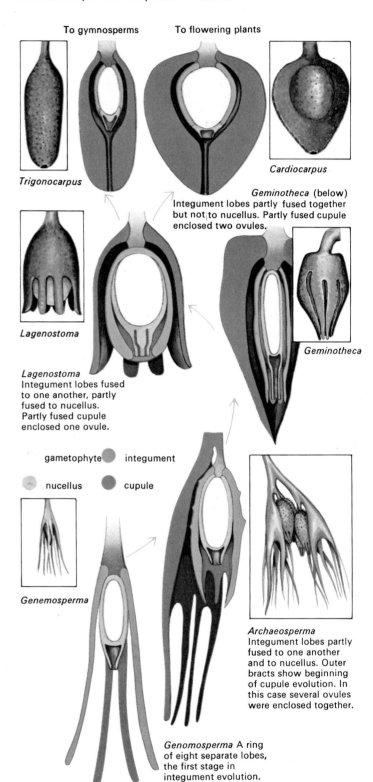

Trigonocarpus Integument and cupule fused together and to the nucellus. Fleshy for animal dispersal.

Cardiocarpus Integument and cupule fused together and to the nucellus.

To gymnosperms   To flowering plants

Trigonocarpus

Cardiocarpus

Geminotheca (below) Integument lobes partly fused together but not to nucellus. Partly fused cupule enclosed two ovules.

Lagenostoma

Lagenostoma Integument lobes fused to one another, partly fused to nucellus. Partly fused cupule enclosed one ovule.

Geminotheca

gametophyte   integument

nucellus   cupule

Genemosperma

Archaeosperma Integument lobes partly fused to one another and to nucellus. Outer bracts show beginning of cupule evolution. In this case several ovules were enclosed together.

Genomosperma A ring of eight separate lobes, the first stage in integument evolution.

Seeds are fertilized ovules. Ovules consist of a single 'female' spore permanently retained within the nucellus. The female gametophyte tissue develops there. In early seeds the nucellus was partly protected by a ring of finger-like lobes. The theory is that the lobes fused together and to the nucellus to provide an extra wall layer, the integument. Some seeds were also partly surrounded by protective bracts. The separate lobes of these fused together into a cupule, evolving into a two-layered integument with a small pore at the tip which allowed pollen to enter.

Fossilized fruits of the seed fern *Caytonia*. In many ways *Caytonia* was more advanced than typical seed ferns and it has been considered to be a close relative of the flowering plants.

Two different kinds of seed fern. *Medullosa*, about 5m tall, grew in the Upper Carboniferous and the Permian. *Caytonia* evolved later, in the Triassic.

Medullosa

Caytonia

actually the parent plant. Other species of *Archaeopteris* were heterosporous and if it can be proved that this particular one did produce seeds, it will be real evidence that seeds evolved from heterosporous reproduction.

More vital evidence comes from fossil plants found in early Carboniferous rocks in Berwickshire, Scotland. Many different types of seeds have been found, showing how variable they were even at this early time, over 300 million years ago.

The most primitive type of seed is *Genomosperma*. Here the nucellus does not have a fully developed protective integument but is partly enclosed by eight finger-like lobes. These probably developed to protect the pollen drop which hung from a special trumpet shaped outgrowth at the tip. In more advanced seeds the lobes have fused to the nucellus as an almost complete protective cover.

These early seeds belonged to a group of plants called the pteridosperms or seed ferns, which survived for millions of years, into the Mesozoic era, before becoming extinct. They seem to have varied from small plants to the 5m tall *Medullosa*, which was superficially very like

Next page: The Wealden region of south-east England about 130 million years ago was a mixture of aquatic, swamp and other communities. Conifers (leafy shoots called *Frenelopsis* and *Pseudofrenelopsis* 1) may have formed forests with cycadeoids (leaves called *Otozamites* 2 and *Pseudocycas* 3), cycads (leaves called *Nilssonia* 4) and tree ferns (*Tempskya* 5). Some smaller ferns like *Weichselia* (6) and horsetails (*Equisetites* 7) were present. Liverworts (*Hepaticites* 8) and clubmosses (*Selaginellites* 9) also grew in suitable habitats.

Dorycordaites

Cordaites and early conifers lived at the same time during the late Carboniferous and early Permian. Some cordaites developed stilt roots similar to those of the mangroves that grow today in coastal swamps. In forests on slightly higher ground other cordaites grew to heights of at least 30m.

Araucarites

Right: Female cone from the Triassic conifer *Voltzia*. As in cordaite cones, the ovule-bearing scales are still quite widely spaced on the central cone stem.

the tree ferns. Others were probably scramblers or climbers. They all had fern-like leaves and for many years the large numbers of Carboniferous seed ferns were thought to be true ferns. It was not until seeds were found attached to the foliage that scientists realized they were something different and the name seed fern was introduced to distinguish them as a group.

Many of the seed ferns developed an important modification of their seeds: a second outer protective layer called a cupule enclosed several seeds together. Like the integument, this began as a ring of finger-like lobes. *Archaeosperma*, the first seed-carrying fossil, had a primitive, divided cupule. Later in the Carboniferous the lobes of the cupule joined together, presumably giving more protection against dessication and animal damage. This development must have been a successful one for Mesozoic forms such as *Caytonia* had cupules that

enclosed the seeds almost completely.

These closed cupules were similar to structures later found in flowering plants and some botanists suggested that seed ferns were the ancestral relatives of the angiosperms. There is good evidence that cycads and a similar group called cycadeoids also evolved from seed ferns. Both cycads and cycadeoids reached a peak of development by the Jurassic period, 190 to 136 million years ago.

Cordaites and conifers

Another group of seed plants evolved from ancestral progymnosperms at the time when seed ferns were diversifying. These were the cordaites, which first appeared in the early Carboniferous, about 325 million years ago. They were tall, with long, strap-shaped leaves arranged in a spiral, amongst which were primi-

Amyelon

tive cones. These plants were significant parts of the vegetation in Europe and North America during the Carboniferous but their importance was not to last. By the mid-Permian they had been replaced by their presumed descendants, the conifers.

The first true conifers came from the later Carboniferous period and differed from their cordaite ancestors because they had more compact, truly cone-like reproductive organs. Rudolf Florin, a Swedish palaeobotanist, studied the fossil remains of numerous Carboniferous and Permian plants, attempting to trace the steps that may have led from cordaite cones to the true conifer cone. After many years of careful detective work he showed that both the male and female cones of cordaites consisted in close up of a series of small shoots, arranged loosely on either side of the cone stem. Each of these shoots bore tiny sterile leaves in a close spiral,

59

amongst which were fertile leaves with pollen sacs (on male cones) or stalked ovules (on female cones). Below each shoot was a scale-like leaf, the bract scale.

Male and female cones probably grew on different parts of the same tree and although both subsequently evolved into conifer cones in a broadly similar way, the changes were by far the most marked in the females. The fertile or ovuliferous shoots of the female cone became progressively smaller and fused together to form a more and more scale-like structure, the ovuliferous scale. The ovule stalks also became shorter so that the ovules became fused to the surface of the ovuliferous scale. At the same time the cone stem became shorter, crowding the scales together into a more compact group and the bract scale became broader to become the outer, protective scale. The loosely packed cordaite cone had its ovules and developing seeds exposed to the air and thus open to attack by predators or to possible drying out. The compact conifer cone on the other hand gave protection which may have been vital for survival, for the humid air of the Carboniferous swamps gave way to much drier conditions in the Permian. The conifers survived but the cordaites vanished forever.

Plant evolution in the southern hemisphere

South of the Equator, plant evolution took rather a different course. Gymnosperms were developing but into distinct forms. Both terrain and climate were different in the southern part of the landmass and there were never the vast swamps that characterized the Carboniferous of Europe and North America. Instead the ground was uneven and there were periods of cold when glaciers developed over large areas. All the plants that grew here in the Carboniferous and Permian were affected by the fluctuating temperatures when cold periods interrupted warmer periods of plant growth.

The southern gymnosperm group, the glossopterids, were dramatically different from their northern relatives. The mixture of leaves that has been found shows that they varied in size from small bushes to large trees. In tropical climates, where there is little variation in temperature during the year, plants shed their leaves gradually and continuously grow new ones to replace them. The fossil leaves from the southern landmass show that plants there shed their leaves all at once at the same time of the year, presumably at the start of a cold season. Such plants are said to be deciduous. The changing climate affected the growth rate of stems as well as leaves and wood formed seasonal growth rings.

From very recent work, the reproductive organs of some of these plants are now known in detail. Petrified

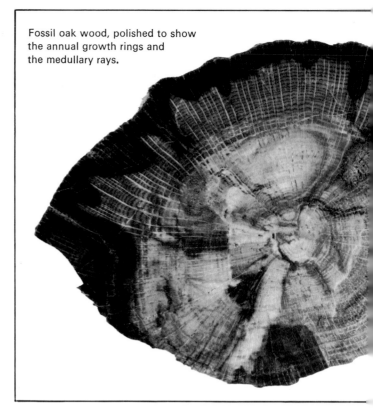

Fossil oak wood, polished to show the annual growth rings and the medullary rays.

cupules found attached to the upper surface of the leaf stalks contained many seeds and now it seems clear that glossopterids were a group of seed-bearing plants with an evolutionary potential equal to any found in the northern hemisphere at the same time. They may even have eventually produced *Caytonia*-like species.

In spite of their early success, by the end of the Permian the glossopterids had disappeared. Once again, we can only guess that a seemingly secure group was wiped out by environmental changes with which they could not cope.

Floral zones

During the 65 million years of the Devonian period many different kinds of plants had appeared and by the end of the Carboniferous, another 65 million years later, most of the major plant groups had evolved: there were clubmosses, ferns and horsetails, seed ferns and cordaites and the first ancestral conifers.

A very important geological change was occurring at this time which had a major effect on the rates of plant evolution. The large landmass which plants had colonized so successfully was changing position and, as this greatly affected climate, the world's plants began to form recognizable groupings now known as Floral Zones.

In the Devonian and the first half of the Carboniferous, the landmass was not yet fragmented into the continents we know today but was divided into two major regions. Laurasia, in the northern hemisphere, consisted of present day North America, Europe and most of Asia. Gondwanaland, in the southern hemisphere, was

Some plants develop secondary thickening as a result of cells in the cambium dividing to form new layers of xylem and phloem. The xylem becomes the woody part of the stem. It continues to carry water and minerals in the sapwood, and medullary rays store food and carry it to the outer areas. As the trunk widens, the central core is compressed into heartwood and no longer acts as a transport system.

The phloem also thickens and continues to distribute food from the leaves. The epidermis, with the soft tissue inside it and usually also old parts of the phloem, become the bark, added to by new cells from the cork cambium.

Annual rings appear when the climate is seasonal and when the tree grows mainly at one time of the year. In the growing season it needs plenty of water and the new wood is composed of large cells. Later on when growth is minimal, the wood is composed mainly of smaller, thicker walled cells and looks darker.

The first fossil wood with annual growth rings is found in Devonian progymnosperms such as *Archaeopteris*, indicating that climates had seasons in the past as well as today.

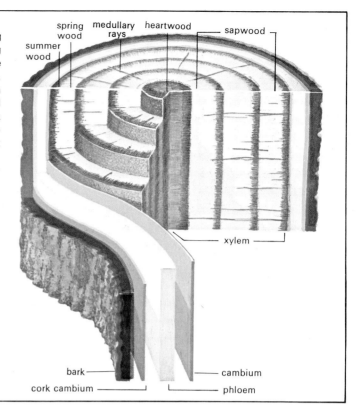

made up of South America, Africa, India, Australasia and Antarctica. Over most of the land the climate was subtropical, warm and moist all the year round.

Because the climate all over the land was similar, there were only minor differences in plant distribution but it is possible to divide plants into two distinct vegetation types or floras. Most of the world had the *Lepidodendropsis* flora, named from the large clubmosses which were so important at the time. Siberia was an exception. Even then its climate was cooler and it had its own Angaran flora, containing plants better adapted to resist the cold.

As plants began to spread into varied habitats in the late Carboniferous the floras began to become more individualistic and distinct and four northern floras and one southern one evolved. The divisions were caused mainly by variations in climate which, together with local differences in topography, controlled the types of plants that were able to grow.

The most luxuriant vegetation was in the Euramerian area where the major coal seams of Europe and eastern North America were being formed by repeated invasions of swamp-forming plants across the vast shallow basins. It was here that the giant clubmosses and horsetails flourished, surrounded by drier uplands covered with early ancestral conifers. This was also the area that suffered most at the end of the Carboniferous period when the swamps dried up, leaving no suitable habitat for the giant plants that had grown there for millions of years.

The North American flora gradually changed throughout the Carboniferous. At first it was similar to

that of the drier parts of the Euramerian flora but it later became strikingly different and so varied that it can be divided into local provinces. In general this was because here plants colonized drier, more upland habitats and then evolved to suit the special conditions of the areas in which they grew.

The Cathaysian flora of China and the Far East was also at first similar to the Euramerian flora although it

Next page: Southern Africa about 280 million years ago was very varied, with glaciers in the mountains and swamp forests in the low-lands. Abundant plants were seed ferns such as *Glossopteris* (1) and *Gangamopteris* (2), growing together with *Noeggerathiopsis* (3). The large clubmoss *Lycopodiopsis* (4) also grew here with small ferns like *Gondwanidium* (5) and horsetails, *Schizoneura* (6).

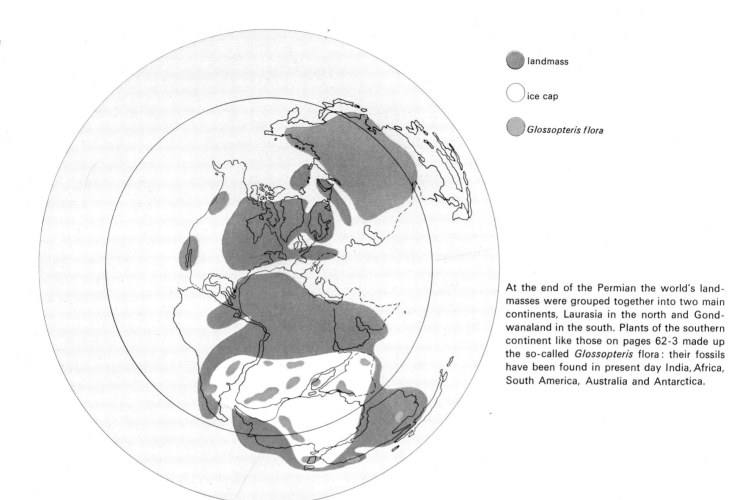

At the end of the Permian the world's land-masses were grouped together into two main continents, Laurasia in the north and Gond-wanaland in the south. Plants of the southern continent like those on pages 62-3 made up the so-called *Glossopteris* flora: their fossils have been found in present day India, Africa, South America, Australia and Antarctica.

already contained several species not found elsewhere. During the Carboniferous it gradually became more and more distinct until by the Permian it was quite separate. The giant clubmosses and horsetails died out there as the climate became drier and there was a steady rise in the numbers and types of gymnosperms. The Pacific Ocean made the climate warmer and the Cathaysian flora was able to spread northwards along the coastal region of China.

Angara was the most northerly of the zones and was too far from the Pacific to be affected by it. None of the Euramerian swamp coal producers grew there, for they needed a warmer and less varied climate. The general Angaran vegetation was therefore much less luxuriant than that of other regions. Although there were club-mosses, they were of a local kind and were small plants and shrubs instead of large trees. There were many kinds of seed ferns and ancestral conifers similar to those of the other northern areas. During the late Carboniferous and early Permian these became more dominant. Different areas of the Angaran flora became more specialized with the northern parts being the most distinct. The other areas changed with time but still shared some species with the Cathaysian and Euramerian floras.

As we have seen, the vegetation of the southern supercontinent, Gondwanaland, became the most dis-tinct of all. In the first half of the Carboniferous glaciers

had covered wide areas of the land, excluding the *Lepidodendropsis* flora which grew farther north. When the ice retreated, whole areas were opened up for coloniza-tion by new forms of plants. The seed fern *Glossopteris* which evolved there soon dominated the vegetation and has given its name to the flora of the whole region.

There is clear evidence in the rocks that advancing ice pushed back the vegetation several times before the plants were able to establish themselves firmly and form large coal producing swamps. Even then the climate was seasonal for many plants were deciduous and grew at variable rates. The *Glossopteris* flora was in the end a successful one as it extended over vast areas, including present day Antarctica. It was also the most stable of all the early floras for it remained almost unchanged until the end of the Permian – for over a hundred million years.

Floral divisions, then, had begun to be established while the landmass remained as one great super-continent. Gradually, they became more and more obviously different until, at the end of the Triassic period (190 million years ago) an entirely new factor came into play. The supercontinent began to break up and the new, smaller landmasses began to move around the surface of the globe, carrying their plants – and animals – into new temperature zones and new en-vironments.

The rise of the seed plants

Conifers and their forerunners the Carboniferous cordaites, existed side by side for some time. By the middle of the Permian, however, the cordaites had vanished, leaving the conifers to establish an ever increasing number of species. Because they are large trees they have dominated the landscape ever since and even today, nearly 300 million years after they first appeared, they cover over 10 million square kilometres, about 8 % of the land surface.

The conifers are not the only group of gymnosperms to have survived to the present day, though they are undoubtedly the most important. Three others, the cycads, maidenhairs and gnetales still have species that are very much alive.

Cycads and cycadeoids

The cycads today are a small family of plants scattered throughout the tropics and of no particular economic importance. They often look like squat palms and only a few develop branches. The tallest may be 18m high but take many hundreds of years to grow to this size. In contrast, some are just a few centimetres tall, with only the tips of the stem showing above the ground.

Cycad leaves grow in a crown at the stem tip and the bases remain on the lengthening stem after the leaves themselves have fallen, forming a tough surface armour. The foliage is very like that of ferns and one genus (*Stangeria*) was even thought to be a fern for many years. However, cycads reproduce like gymnosperms: indi-

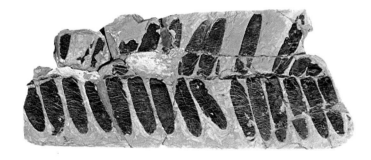

Left : *Otozamites*, the leaf of a large cycadeoid plant. Cycadeoids were alive for about 100 million years, from the Triassic to the Cretaceous period and were probably as common during this time as the true cycads. Today, they have completely disappeared.

The gymnosperms evolved from primitive seed-bearing relatives of the progymnosperms and include the now extinct seed ferns. Four main groups are still living. Conifers are numerically the most successful — second only to the flowering plants. Cycads and ginkgos were in the past much more numerous and varied, particularly in the Jurassic. The fourth living group, the gnetales are unknown in the fossil record.

Growing tip of a female *Cycas revoluta,* encircled by a zone of brown fertile leaves with ovules.

Quaternary	2	Cycads	Ginkgos		Conifers
Tertiary	64				
Cretaceous	136	Cycadeoids	Seed ferns	Caytoniales	
Jurassic	190				
Triassic	225				
Permian	280	Cordaites			Glossopterids
Carboniferous	345				
Devonian	410 million years ago	Progymnosperms		no fossil evidence	

vidual plants produce either male or female reproductive organs and the female ones have primitive ovules. These are large and have archegonia which are fertilized by swimming sperm liberated from pollen grains.

Cycads evolved from seed ferns in the Upper Carboniferous and at first were so like seed ferns that it is impossible to know where one group ended and the other began. The early forms had simple, undivided leaves with a crown of fertile leaves on which grew their male or female reproductive organs. Once the fertile leaves had appeared the stems stopped growing and because this meant the plants could only reproduce once in their lifetime, later species developed methods of continuing growth. Some, like the living *Cycas*, grew alternating sterile and fertile leaves; others, like living *Zamia* and *Macrozamia*, formed cones every few years. If a plant takes many years to grow to maturity and then has only one chance to reproduce, this is obviously wasteful. If each one can reproduce many times during its lifetime, the group as a whole has a much better chance of surviving.

The early origin of the cycads makes them rival the conifers as the oldest types of seed plants still surviving today. They continued to evolve until the Jurassic (190 to 136 million years ago) when the species were apparently as advanced as living ones today. However, since then the group has gone into a decline. Many important genera have become extinct and those that survive contain far fewer species. If the cycads had become extinct and we knew of them only from fossils, they would be accepted as a once successful but primitive group that could not adapt sufficiently to survive. The fact that some are still alive shows how apparently primitive plants can exist even in competition with many more advanced rivals.

Vast numbers of other cycad-like plants have been found as fossils in Jurassic rocks. These look superficially like the cycads but they have different stem and leaf anatomy and different reproductive organs. These cycad-like plants, called cycadeoids, are interesting because the best known ones had reproductive organs rather like flowers and for many years they were thought to be forerunners of the flowering plants. However, we now know that they were not in the direct evolutionary line. Their 'flowers' were mainly clusters of male and female reproductive organs which were protected by up to a hundred hairy scales. In some species as many as five hundred 'flowers' appeared at the same time on the stem, suggesting that they may have reproduced only once in their lives. Like the cycads, the cycadeoides developed from the seed ferns and were at their most varied in the Jurassic. They then went into a decline and, by the end of the Cretaceous, 64 million years ago, they had become extinct.

Palaeocycas

Early cycad fossils like *Palaeocycas* provide rare glimpses into cycad evolution. This well known reconstruction is very imaginative, for the stem, seeds and the way the leaves and cone scales were borne on the plant are quite unknown. *Williamsonia*, one of the many cycad-like plants from the Jurassic, is much more fully known as is the Cretaceous true cycadeoid *Cycadeoidea* which had clusters of male and female organs on its bulbous stem.

Cycas rumphii. Cycads reached their peak in the Jurassic, when they grew as far north as present day Greenland and Siberia. Today they are restricted to tropical and subtropical regions.

The maidenhairs

Maidenhairs, like cycads, belong to a very old group of plants which is now long past its peak. The Maidenhair tree (*Ginkgo biloba*) is the only species still surviving and today has a very limited wild distribution in the forests of western China. It has also been kept as a sacred tree in temple gardens in China and Japan and is grown for its edible seeds which are thought to help digestion and reduce the effects of alcohol.

Ginkgos are tall, graceful trees up to 30m high, with dainty, pale green foliage which turns golden yellow before being shed in the autumn. Their characteristic fan shaped leaves look like the leaflets of the maidenhair fern, although they are larger and much more leathery in texture. As in the cycads there are separate male and female plants and they have similar reproductive systems: pollen grains from the male plant arrive at the ovules of the female and liberate sperm which swim in the pollen drop liquid to the archegonia.

Ginkgos are hardy and are easily and often grown as ornamental trees. They were introduced into Europe at the beginning of the eighteenth century and can now be seen growing at roadsides, in parks and gardens in many temperate regions of the world. The fact that they grow so easily in places where they are no longer found in the wild suggests that they were once much more widespread and this is borne out by their fossil history.

Williamsonia

Cycadeoidea

An impression fossil of the underside of a cycadeoid 'flower' of the Jurassic period. Nectaries on the petal-like scales may have been attractive to insects, preparing them for the evolution of true flowers in the succeeding Cretaceous period.

The first recognizable maidenhairs lived in the Triassic period (225 to 190 million years ago) and by the next period, the Jurassic, they had virtually worldwide distribution. This was their climax; although they continued to live in many places in North America, Europe and Asia until the Tertiary began 64 million years ago, they were gradually becoming less numerous. They were always very common in north-eastern Siberia and this probably saved them from extinction for there the effects of climatic change and competition from other plants were less. Late in the Tertiary they spread southwards to China, where they still grow today.

The conifers

Conifers share certain characteristics with the cycads and ginkgos which enable us to group them all as gymnosperms but they also have other features which set them apart as more advanced and successful forms. Like the ginkgos they are trees, divided into a characteristic woody tree trunk and a branching crown. Although their growth form seems to have been more adaptable, allowing them to survive in a wide range of environments, it is their reproductive system which has probably given them so much success.

Male and female cones commonly grow on the same plant, varying in size and shape according to species. Conifer pollen lands on the ovule in the usual way and grows a pollen tube. Instead of releasing swimming sperm, however, it discharges male nuclei directly into the egg cell. This is a very reliable method of fusion. The ovules are relatively small until they are fertilized so food does not have to be wastefully transferred from the parent plant and stored in ovules that may never become seeds. No such saving occurs in the more primitive gymnosperms, which automatically form full-sized ovules before pollination has even occurred.

Conifer reproductive organs today are nearly always clustered into male and female cones although the yew (*Taxus*) and some podocarps (*Podocarpus*) produce single seeds on small specialized shoots. The male cones usually fall soon after the pollen is released into the wind and are too delicately built to survive for long. In contrast, the female cones are robust and usually become woody and overwinter before shedding their seeds. The pines take this overwintering even further and do not drop their seeds until the second growing season after pollination. This is probably an adaptation brought about as pines colonized areas with short growing seasons. During the long winters there would be little chance of producing food by photosynthesis so it would be impractical to use food reserves for seed

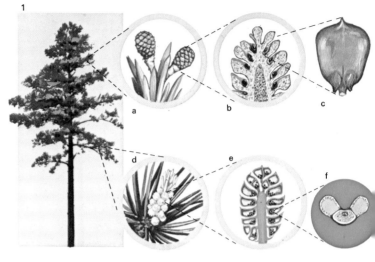

Life cycle of the conifer *Pinus*

dark blue = haploid   light blue = diploid

Conifer reproduction is more efficient than that of more ancient seed plants like cycads. Pollen is drawn into the ovule in a similar way and a pollen tube grows towards the egg. However, sperm cells no longer swim at the final stage but are discharged directly into the egg, without swimming, so that fertilization is almost certain to take place.

A 60 million year old fossil *Ginkgo* leaf from Wyoming, USA with (below) new spring leaves and male catkins of the only living species, *Ginkgo biloba*.

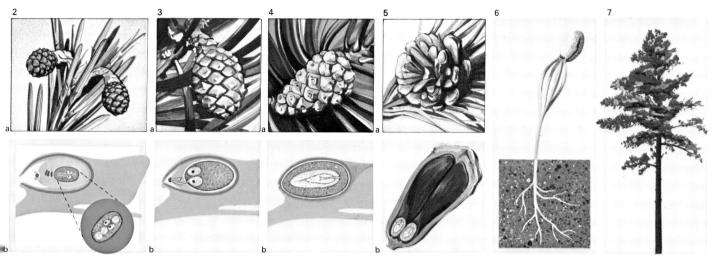

1 Male and female cones grow on the sporophyte tree. The female cone (a) contains ovules (b) arranged on the scales (c). The male cone (d) has pollen sacs (e) which produce pollen grains (f). In spring the female cones are open to receive pollen. 2 One month after pollination the female cone (a) has closed again. Inside the ovule (b) the pollen grain has been drawn in and the female gametophyte which will produce the eggs is just beginning to grow. 3 Fifteen months later the cone has enlarged (a). Inside the ovule (b) the female game-tophyte has produced its egg and the pollen tube has discharged sperm cells into it. 4 Almost two years later the cone has reached full size but is still tightly closed (a). Inside the ovule (b) the embryo seedling has developed. 5 Two years after pollination, the cone opens to allow the seeds to fall out (a). Each scale (b) contains two ripe seeds, with wings formed from a thin layer of the scale. 6 The new seedling begins to grow. 7 When the tree is mature, the cycle begins again.

growth. Spreading seed growth into another growing season would have had obvious advantages for the parent plant.

The jack pine (*Pinus banksiana*) keeps its cones closed until forest fire has passed over them, an amazing adaptation to a very difficult environmental factor. Forest fires often occur spontaneously in dry vegetation and can spread over vast areas, destroying everything in their path. Any plant that protects its seeds against fire and then liberates them in vast numbers to recolonize the devastated area will be very successful there.

Evolution has gradually changed the form and reproductive features of the conifers to produce a very mixed group of modern families and genera. Change has not of course proceeded at the same rate in all parts of the plant and all families have both primitive and advanced characteristics which can be compared and which help to build up a picture of what the original forms were like.

The evolution of the conifers coincided with a very important geological time which involved the breaking up of the continents. This splitting had possibly begun as early as the Triassic but wind-borne seeds and spores could still be carried from one landmass to another until well into the Cretaceous period. Separation was at its most extreme during the late Cretaceous and early Tertiary for the continents had not yet formed their new links. In addition, much present day land was covered by shallow seas where thick sedimentary deposits were being laid down. South America was nearly cut into two by the immense Amazon Basin Sea and Eurasia was split by the great Tethys Sea which joined the Arctic and Indian Oceans. In many isolated areas plants made special adaptations which were to mix when the continents came into their familiar positions in the Tertiary period.

The wandering continents carried their floras with them into new climatic zones while the continents themselves were being altered by the movements. Mountain ranges such as the Rockies, the Andes and the Himalayas were being pushed up and were altering rainfall patterns and climatic conditions over thousands of square kilometres. It is hardly surprising that this was a time of great evolutionary change among both plants and animals.

By the end of the Tertiary, 2 million years ago, conifers had evolved into the plants we know today in the temperate regions of the world and all later changes in the fossil records merely show plant migrations and extinctions.

Obviously when botanists talk about plant movements or migrations they do not mean that individual plants move, rather that the distribution of the species as a whole changes. Changes in climate can alter the area in which the species can survive, reproduce and establish its seedlings: new areas may become suitable for colonization while others are becoming uninhabitable. Seeds and spores carried into new, suitable areas will become established while those that remain in the original environment will fail to grow. Providing the changes are slow enough, plants will be able to colonize the new areas fast enough to avoid extinction.

As the great Ice Ages approached and the climate in the northern hemisphere cooled, plants moved gradually southwards. Conifers that had grown in the Arctic regions during the warmer Tertiary period now migrated through North America, Europe and Asia, spreading across three continents. Often species were isolated in limited areas, though the genus as a whole may still have been widespread. The effect of the colder climate was much more drastic in Europe than it was elsewhere and fossil floras show a record of widespread extinctions throughout the area. Mountain ranges in southern Europe and around the Mediterranean made it impossible for plants to spread further south and waves of migratory plants were trapped in unsuitable climates and wiped out. During the late Tertiary many genera and even families of conifers disappeared from Europe.

Today there are generally accepted to be seven living families of conifers, divided into 54 genera and 570 species. Most had many more close relatives in the past but although they are now mainly restricted to the northern hemisphere, they are still a very varied and interesting group, ranging from the 112m high coastal redwoods to dwarf arctic and alpine pines and junipers. They also include the longest lived trees, the bristlecone pines (*Pinus aristata*). Most are evergreen, although

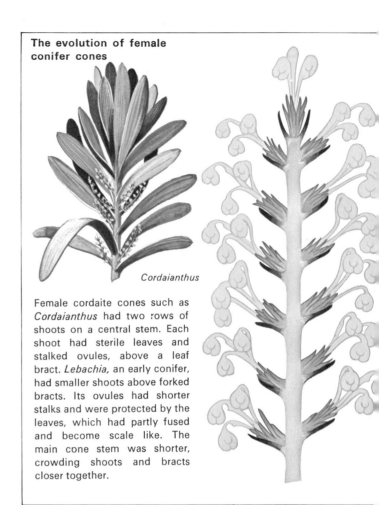

The evolution of female conifer cones

Cordaianthus

Female cordaite cones such as *Cordaianthus* had two rows of shoots on a central stem. Each shoot had sterile leaves and stalked ovules, above a leaf bract. *Lebachia*, an early conifer, had smaller shoots above forked bracts. Its ovules had shorter stalks and were protected by the leaves, which had partly fused and become scale like. The main cone stem was shorter, crowding shoots and bracts closer together.

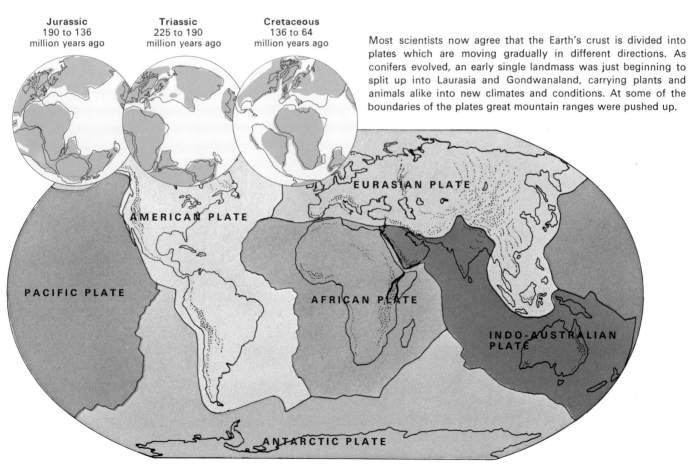

Jurassic
190 to 136
million years ago

Triassic
225 to 190
million years ago

Cretaceous
136 to 64
million years ago

Most scientists now agree that the Earth's crust is divided into plates which are moving gradually in different directions. As conifers evolved, an early single landmass was just beginning to split up into Laurasia and Gondwanaland, carrying plants and animals alike into new climates and conditions. At some of the boundaries of the plates great mountain ranges were pushed up.

EURASIAN PLATE

AMERICAN PLATE

PACIFIC PLATE

AFRICAN PLATE

INDO-AUSTRALIAN PLATE

ANTARCTIC PLATE

Lebachia

Abies alba

A modern conifer cone such as that of the silver fir (*Abies alba*) is very compact. The leaf scales of the fertile shoot have completely fused together and, with the ovules fused to them, they form the ovuliferous scale. The bract scale has become larger to enclose and protect the developing ovules.

Fossil cone of *Pinus florissanti*, a pine which grew in Colorado during the Oligocene.

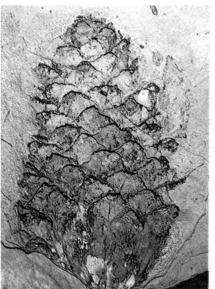

cone stem

ovuliferous
scale

ovule

bract scale

some, like the larch (*Larix*) shed their leaves every autumn and the swamp cypress (*Taxodium*) and dawn redwood (*Metasequoia*) actually shed small shoots.

MONKEY PUZZLES AND KAURIS

These two groups make up a very ancient family (Araucariaceae) which has been recorded from as far back as the Triassic period in both hemispheres.

Monkey puzzles (*Araucaria araucana*) are perhaps the most primitive of all living conifers and they must surely rate as the most improbable looking trees of all time. Their straight trunks with the rings of stiff horizontal branches and regular pairs of side branches give them a precise shape unique in the plant world. The regularity is accentuated even more by spirally arranged, large, dark green leathery leaves which may live for fifteen years and remain on the tree for many more before they rot at the base and finally fall.

Monkey puzzles grow wild today in the Andes of Chile and Argentina and have close relations in Australia, New Guinea and New Caledonia. In the Triassic when they evolved all these countries were joined together so it is not surprising that they now grow on different sides of the world. They are extremely successful in the Andes where they often form pure forests, excluding all other trees. Fossils from the Cerro Cuadrado petrified forest in Patagonia show that they

Next page: A scene in south-western North America about 200 million years ago. Conifers with leafy twigs called *Brachyphyllum* or *Pagiophyllum* (1) and monkey puzzles like *Araucarites* (2) dominated the humid, warm landscape. Large cycadeoids (with *Otozamites* leaves 3) and tree ferns (*Itopsidema* 4) were conspicuous while ferns like *Cladophlebis* (5), *Cynopteris* (6), *Phlebopteris* (7) and *Todites* (8) also occurred. There were many horsetails (*Equisetum* 9), probably growing along river banks and in damp places.

have been in the area since the late Jurassic, for nearly 200 million years.

The two other important species in this genus are the delicate Norfolk Island Pine (*Araucaria heterophylla*) from the Pacific and the bunya-bunya (*A. bidwilli*) from Australia. The bunya-bunya is of great economic importance to the Australian aborigines. Like all monkey puzzles bunya-bunyas produce large female cones, in this case sometimes as large as footballs. Each cone produces up to 150 large seeds which are highly prized as a valuable food crop.

The kauris (*Agathis*) of New Zealand are grouped with the monkey puzzles because they have similar reproductive systems but they grow in a rather different and in many ways a more impressive way. Some are over 50m tall, with smooth unbranched trunks for well over half their height, topped by irregular crowns of branches. This makes them a very useful timber crop and they are also a source of the copal resin used in the paint and linoleum industries.

REDWOODS AND CYPRESSES

The redwoods and their relatives make up the family which shows most clearly that conifers were once much more widespread trees. Now surviving in four areas, one group on each side of North America, one in China and one in Tasmania, their fossil record proves that they originally played important roles in the forests which covered the northern hemisphere.

The family (Taxodiaceae) is known from Jurassic fossils over 130 million years old but reached its climax in the Tertiary. Climatic conditions then were very much warmer than they are today and in many areas swamps formed once more. This time, however, instead of clubmosses and horsetails, conifers were the important trees and large redwoods dominated the vegetation for thousands of square kilometres. The thick deposits of brown coal which is so important in parts of Europe were formed from the remains of these swamp forests in the same way as earlier coal deposits formed from clubmosses and horsetails in the Carboniferous.

Coal bearing swamps remained for over 60 million years until the climate began to cool and the Ice Ages approached. During the Ice Ages the family was reduced to 15 species, all, except the three Tasmanian *Athrotaxus*, in the northern hemisphere.

The best known of the survivors is the magnificent redwood *Sequoia sempervirens* which is now limited to the coastal belt of the south-western United States. The big tree or wellingtonia (*Sequoiadendron giganteum*) is even more restricted, growing only in the high valleys of

Sierra Nevada in central California. These immense trees reach over 100m in height, 30m in circumference and 2,000 years in age; both they and the coastal redwoods have very thick spongy bark and are well protected against the raging forest fires that often sweep through their homelands.

The swamp cypress (*Taxodium distichum*) was once an important tree in the Tertiary swamps but is now found only in the swamps of south-eastern United States. Like its close relative the dawn redwood (*Metasequoia glyptostroboides*) it loses its leaves in winter. The dawn redwood is unusual in having been known longer as a fossil than as a living tree. It was first described as a fossil in 1941 but it was not until 1944 that it was found growing wild in China.

True cypresses (Cupressaceae) are evergreen, ranging in size from shrubs to large trees. Fossils are known from the Jurassic and today they are widespread with species in Africa, Europe, North America, Asia and Australasia.

THE PINES

The pine family (Pinaceae) includes cedars, spruces, firs and larches as well as true pines. *Pinus* itself is the largest genus of all the conifers and is divided into 90 species. They range from the giant 75m tall sugar pine (*Pinus lambertiana*) to smaller forms such as *Pinus*

Kauri pines (*Agathis australis,* right) and monkey puzzles (*Araucaria araucana,* left) are grouped in the same primitive family, the Araucariaceae, which has existed since the Triassic. Both are mainly southern hemisphere trees, kauris growing from the Philippines to New Zealand and monkey puzzles in South America, Australia, New Guinea and New Caledonia.

Conifers are usually divided into seven living families. Like many other gymnosperms they were more numerous in the Jurassic and Triassic, when several extinct families were alive.

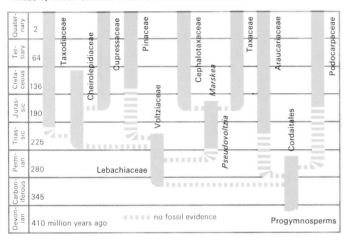

mugo which grows on mountain screes high up in the Swiss Alps and Pyrenees. They can be distinguished from all other conifers by their needle-like leaves which are arranged in bundles of two, three or five.

The pine family probably existed in the Jurassic period, although there are no completely convincing records. The oldest living genus, *Pinus*, is known from Belgium in the lower Cretaceous but most modern species probably originated within the last two million years.

Today pines are very successful trees, growing in the coniferous forests which cover vast areas of land between 40° and 70° north of the equator. They either dominate large regions or grow mixed in with spruces, firs and larches. The exact species of conifers of course changes from place to place, reflecting the independent evolution that occurred when plants migrated southwards during the Ice Ages.

Many pines live for hundreds of years but the bristlecone pine *Pinus aristata* holds the record, with some trees of nearly 5,000 years old. They live in very harsh climatic conditions in the White Mountains of California where growth often seems impossible: yet they survive. They must have colonized the area when the climate was much milder and are now merely survivors of a past age. The main reason they are able to live for so

long is that their growth rate is extremely slow so that they are in a state of near suspended animation. Even so, they are largely dead stumps of contorted wood with very few living branches. Life is made possible only by narrow strips of living bark which connect the leafy branches with the root system. Only a few trees remain and it seems likely that they will soon become extinct.

Another pine, *Pinus strobus*, the white pine, once dominated much of northern North America. However, it has been over exploited as a timber tree and now grows wild in only a very small area. The Douglas fir (*Pseudotsuga*) has become the principal lumber tree of the area but as forests of these are replanted as part of the timber management plan, they are unlikely to become rare.

Christmas trees, probably the best known of conifers, belong to a genus that first appeared in North America in the Upper Cretaceous, between 100 and 64 million years ago. They are Norway spruces (*Picea abies*), one of many species of spruce that now grow well all around the northern hemisphere. They grow in a regular triangular shape when they are small but lose their lower branches as they grow, especially if they are close together. In ideal conditions mature trees may reach 60m high and they form an important part of the dense conifer forests of the north.

Firs (*Abies*) appeared early in the Tertiary period (38-26 million years ago). All northern plants were forced to migrate southwards during the Ice Age but firs, like other conifers, were well adapted to withstand cold, dry climates. Their tough, needle-like leaves prevented too much water loss while their straight, tall stems and regularly arranged, drooping branches enabled them to withstand strong winds and to carry heavy weights of snow. When the ice retreated they gradually migrated north again to recolonize the land.

Spruces and firs are often confused but there are two quite obvious ways of telling them apart. Like almost all conifers, they are evergreen, replacing their leaves gradually instead of all at once. When old spruce leaves fall they leave behind small woody pegs which give the branches a knobbly look; in the firs, however, the old needles leave rounded scars almost level with the stem surface.

The second distinguishing feature involves the female cones and again is very noticeable. Spruces have cigar shaped female cones which hang downwards and open to release their seeds. Female fir cones stand upright and the winged seeds fall out as the cones slowly disintegrate.

The Douglas fir, which is such an important timber tree in North America, belongs to a different genus from the true firs. It grew in Europe during the Tertiary but died out there in the Ice Age and is now found only in North America and eastern Asia.

Larches evolved in the Tertiary. They are more closely related to true pines than to firs and spruces but are included here because they are similar in general appearance. All twelve living species grow into pyramid shaped trees with slender needle-like leaves. Unlike almost all other conifers, larches are deciduous, dropping their leaves in winter and producing a new crop in the spring. We do not know why they evolved this characteristic but it was presumably to enable them to survive in a relatively dormant state throughout the time of year when the climate was particularly harsh. Today they in fact grow as far north as any of the hardiest evergreens.

Like the redwoods, cedars now remain only in small, isolated areas but they first appeared in the Tertiary and were once very widespread throughout Europe. Now there are only four species, growing wild in widely separated parts of the world, all in mountainous regions. The smallest, *Cedrus brevifolia*, comes from Cyprus and the largest, the Indian or deodar cedar (*Cedrus deodara*) grows in the western Himalayas. The Atlas cedar (*Cedrus atlantica*) is found in the Atlas

Bristle-cone pines (*Pinus aristata*), high in the Rocky mountains, are the longest lived of all trees, some nearly 5,000 years old. They grow extremely slowly and have very few live leafy branches. They survive because their dead wood does not rot in the dry atmosphere.

Right: The wellingtonia or 'big tree' (*Sequoiadendron giganteum*) grows on the western slopes of the Californian Sierra Nevada. It is one of the biggest trees in the world, some standing over 100m tall.

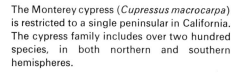

The Monterey cypress (*Cupressus macrocarpa*) is restricted to a single peninsular in California. The cypress family includes over two hundred species, in both northern and southern hemispheres.

Swamp cypresses (*Taxodium distichum*) are amongst the few species of conifers that are deciduous, losing their leaves and small branches in autumn. This species is now found naturally only in the swamps of southern-eastern United States.

mountains of North Africa while the cedar of Lebanon (*Cedrus libani*) now grows wild in only a few small groves in the mountains of Lebanon. The cedar of Lebanon has, however, a much wider cultivated range. Because of its natural beauty it is grown around the world as an ornamental tree and is a familiar plant in the grounds of European stately homes.

YEWS, PLUM YEWS AND PODOCARPS

The three families of yews (Taxaceae), plum or head yews (Cephalotaxaceae) and podocarps (Podocarpaceae) are rather different from other conifers for most do not have cones. Instead all yews and many podocarps have single seeds which are completely or partly enclosed by red or crimson fleshy coats and look like berries. The fleshy coats attract birds which eat the seeds and later disperse them over wide areas, ensuring a good distribution for the species.

Both yews and podocarps can be traced back in the fossil record to the Jurassic period, the yews in southern Poland and northern Britain and the podocarps in India. Yews are still found mainly in more northerly areas with podocarps further south – in India, Malaysia, Japan and parts of China. Plum yews are often cultivated in Europe as decorative trees but grow wild from the Himalayas to Japan.

The wide distribution of yews and podocarps shows that they probably evolved before continental drift

A spruce forest (*Picea*) in Alberta, Canada. Dense conifer forests grow in the far north, the trees crowding together so that they shade the ground and make it difficult for many other species to grow among them.

Right: Outeniqua yellow-wood (*Podocarpus falcatus*), a tall podocarp growing in Tsitsikama forest, South Africa. Some podocarps have leaves more like broad-leaved trees than conifers.

split up the southern continent of Gondawanaland. There is some evidence that they invaded the south-eastern USA in the Tertiary but later died out there.

Conifers became successful plants a very long time ago. Superficially they seem almost unchanged by over 60 million years of evolution but during that time they have made many important adaptations which have enabled them to survive and flourish in changing climates and environments. If man allows, they will continue into the future, dominating some areas and sharing many more with the flowering plants.

The gnetales

The last group of gymnosperms, the Gnetales, have no commercial value and are of no use as ornamental plants so most people know little about them and they have no widespread common names. Scientifically they have been considered alternatively as immensely important to our understanding of plant evolution and the origins of flowering plants – and as mere botanical curiosities.

The three genera are grouped together because they have characteristics which are more like those of flowering plants than those of gymnosperms. Their reproductive organs are arranged in male or female groups with scales around the male 'flowers' and ovules rather like the sepals and petals of true flowers. Also like flowering plants, they have specialized conducting cells in their wood called vessels. Some botanists believe they are ancestral flowering plants but the most likely explanation is that they evolved these characteristics independently to suit their particular environment. They are in fact examples of parallel evolution which is said to have taken place when different plants or animals make similar adaptations quite separately.

Two of the genera (*Ephedra* and *Gnetum*) are very

Welwitschia mirabilis, one of the Gnetales, is a desert plant from south-west Africa. Its two long leaves last throughout its life (for up to two thousand years), growing from the base as their tips fray and are destroyed. The large storage root may grow up to 1 m in diameter.

Above, left : Yews have single ovules at the tips of small shoots. When the seed is developing, a fleshy edible outer coat grows from the seed stalk, attracting birds to disperse it. The seeds are poisonous.

Left : *Ephedra fragilis* comes from the coastal Mediterranean. Ephedras are mainly small, shrubby plants, sometimes climbing or trailing. Their leaves are tiny : the stem carries out most of the photosynthesis.

widespread and vary in size and form from shrubs to trees and even to tropical creepers. The third genus has only one species (*Welwitschia mirabilis*) which lives in one of the smallest and oldest deserts in the world, the Namib desert in Namibia.

Welwitschia mirabilis is among the most curious of plants living today. They survive for up to 2,000 years but still have the same basic shape of young plants. Each has a woody turnip shaped stem with a single pair of leaves which are the most conspicuous part of the plant. They grow continually from the base and if they die the whole plant dies too. With such a long life, the leaves can reach lengths up to 9m, although they become torn and tangled over the years.

Many desert plants survive because they have an enlarged root system which absorbs any available water. *Welwitschia* depends mainly on its leaves. Heavy fogs shroud the coastal regions of Namibia for most days

in the year and these are often carried over the desert by westerly winds. The fog condenses on the leaves as water and is absorbed through the vast numbers of stomata to be rapidly conducted to other parts of the plant. When the fog lifts, the stomata close, cutting down water loss during the heat of the day.

A great deal has been written about the relationship of the Gnetales to flowering plants and to each other. They do have many similarities to flowering plants but there are also important differences in the way they grow and are fertilized. There is very little fossil evidence for any of the three genera : the only fossils of *Ephedra* are pollen grains from the Tertiary period and possible other remains from the earlier Permian. It seems likely that the group is an artificial one and that the three types evolved from different ancestors. It is a group that, more than many others, shows how little we really know about the evolutionary history of plants.

The flowering plants appear

The angiosperms or flowering plants are the most successful group of plants that have ever lived. There are about 250,000 living species compared with about 50,000 species of all the other green plants put together. They have an enormous range of size, form and structure and are able to colonize most habitats in the world where life is possible at all. The angiosperms are more productive than other plants and supply the most successful groups of animals, the mammals and insects, with nearly all their food. It is not surprising to find that these animals expanded and diversified at the same geological time as the angiosperms. The Russian botanist Takhtajan has put it more dramatically but equally correctly by saying that the dominance of flowering plants made possible the appearance of man.

To find the beginnings of the flowering plants we must once again go back in time. Just as the gymnosperms took over only gradually from the seedless clubmosses and horsetails, so the angiosperms faced a similarly difficult start to life in a world dominated by long lived and well adapted forest species.

The dramatic success of the angiosperms in the Cretaceous period (136 to 64 million years ago) completely altered the landscape. Until about 112 million years ago the dominant land plants were gymnosperms but within 10 million years many angiosperm families had appeared and by the end of the Cretaceous most major living groups were recognizable. But what made them so successful? It is too simple to say that a flower by itself makes a plant more competitive and successful and in fact there are several features in addition to flowers which distinguish the average angiosperm from the average gymnosperm. As usual when dealing with living organisms, there are exceptions.

What is a flower?

The flower of a flowering plant is its reproductive organ, corresponding in function to the sporangia of clubmosses and the cones of gymnosperms. It is quite easy to recognize a flower but virtually impossible to define exactly what it is; there are far too many different types for any one simple description to apply to them all.

The average picture of a flower shows it to have protective sepals and attractive coloured petals (together called the perianth) with pollen-producing stamens and ovule-bearing structures called carpels. Some flowers have no perianth and others are either male or female, having only stamens or only carpels. Stamens are very like the structures found in the Gnetales so they cannot be used alone to define a flower. Ovules and carpels remain as some of the distinctive features that separate most angiosperms from all other plants.

There are many different types of female reproductive structures in angiosperms and by comparing them we can trace the way they have evolved. In some the carpels are like folded leaves, with ovules attached to their inside

Life cycle of an angiosperm

dark blue = haploid
light blue = diploid

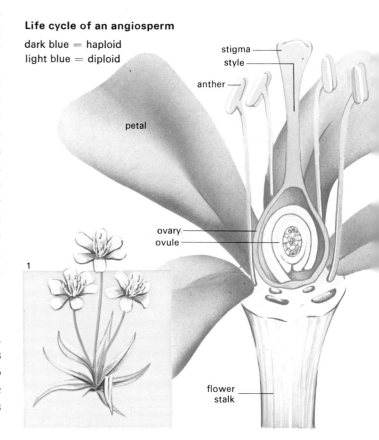

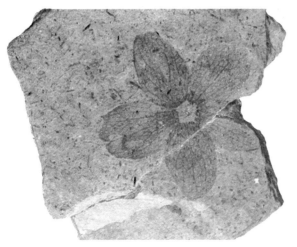

A fossil flower *Porana oeningensis*, thought to be a member of the Convolvulaceae. It was found in late Tertiary deposits in Oeningen in Germany; other specimens are known from Switzerland.

surfaces. The gap between the carpel lips is tightly closed by densely packed hairs. It is thought that all early angiosperm flowers were like this. Later, the carpel lips fused together, closing the gap completely. This isolated the ovules from the outside world, protecting them from drying up, fungal infection and insects.

Apart from the fact that they are enclosed, there are no immediately obvious visible differences between the ovules of gymnosperms and those of flowering plants. It is true that most angiosperm ovules have two protective integuments while all gymnosperms have only one, but it is inside the ovule, in the tissue itself (the prothallus) that most change has taken place. Under a microscope it is clear that the prothallus of an angiosperm is much less complex than that of a gymnosperm. A gymnosperm prothallus is multicellular and develops sex organs (the archegonia) at its specialized tip. Angiosperm prothalli, unlike nearly all other vascular plants, do not have archegonia and remain very small, usually with only seven cells, six with single nuclei and one with two nuclei.

Angiosperms can, therefore, obtain the same result as the gymnosperms but more efficiently. The prothallus

Flowers are the reproductive organs of angiosperms. Pollen is produced by the anthers and the central female ovary contains the ovules. Anthers and ovary are often surrounded by attractive petals which guide pollinating insects to the right place.

1 Mature plant produces flower. 2 Anthers (a) produce pollen grains. Inside the ovule (b) is a small, seven-celled prothallus. 3 Pollen grains trapped on the stigma grow tubes through the gap in the integument (micropyle) to the tip of the nucellus. The tube passes through the nucellus and embryo sac into the prothallus. 4 The two male nuclei carried in the pollen tube fuse with nuclei of the prothallus. 5 One fusion pair grows into the embryo. The other pair forms food reserves. 6 The seed matures and is dispersed when the ovary wall splits open. 7 The seed germinates. When the new plant is mature it flowers and the cycle begins again.

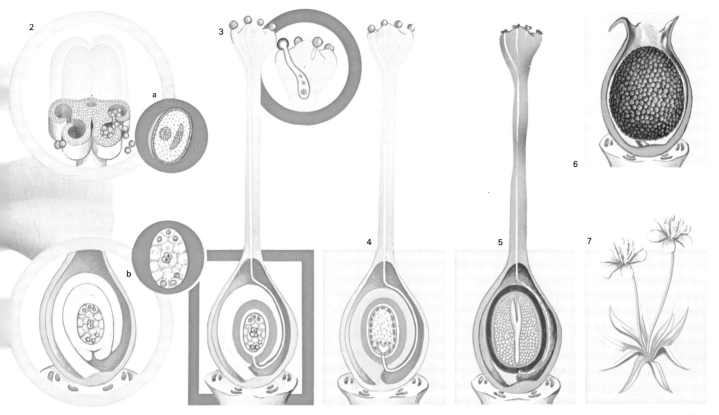

consists of only a few cells, even smaller than a conifer ovule. It remains like this until fertilization so there is no wastage if it is not fertilized.

Pollen grains, the male reproductive structures, also changed in the flowering plants. These changes are directly related to the development of closed carpels which at first sight appear to seal off the female egg not only from outside dangers but also from the fertilizing male sperm. In the early plants the male spores or the pollen grains liberated swimming sperm which attempted to make their way to nearby archegonia with varying degrees of success. Two groups of gymnosperms, the conifers and gnetales are exceptions for instead of releasing swimming sperm their pollen tubes grow right into the egg cell to fertilize it. This is essentially what happens in angiosperms where pollen tubes carry the sperm nuclei to the egg nuclei within the ovules.

Because the ovules are now hidden away within the carpels a special receptive area called the stigmatic surface or simply the stigma evolved on the outside. Pollen grains landing on this surface begin to grow a pollen tube which at this point is able to penetrate the carpel wall and extend to the ovules, usually through a gap at the tip of the protective integument. The pollen tube then forces its way through the ovule wall and into the embryo sac.

The events which follow are also of great importance. A free swimming sperm has a single nucleus, but an angiosperm's pollen tube has two nuclei at its tip which both enter the embryo sac together. One of these fuses with the egg cell nucleus in the usual way to become the diploid cell which will produce the embryo and grow into the next plant. The second nucleus in the pollen tube, however, fuses with either another one or another two in the embryo sac (remember there are usually eight) and later divides many times to become food reserve tissue surrounding the developing embryo. Such a process is often called double fertilization and probably occurs in all angiosperms.

Another important feature found in angiosperms is connected not with their reproductive system but with the phloem, the system by which food is carried from the leaves to all parts of the plant. In angiosperms the conducting cells are joined end to end into tubes and there are specialized cells called companion cells which help in the active transport mechanism. This more complex system is presumably more efficient although it has never been experimentally proved.

The transport system, the small number of cells in the embryo sac and the double fertilization process are the really important characteristics of angiosperms. The chances of these three complex features developing simultaneously in more than one group of plants seems very unlikely and most botanists therefore believe that all angiosperms evolved from a common ancestor.

The origin of the angiosperms

'That abominable mystery' was how Charles Darwin described the evolutionary origin of the angiosperms. Their sudden and almost inexplicable appearance and their equally rapid rise to superiority are facts, proved by studying fossil plants from the Cretaceous period. But why they evolved and where they came from are questions that have been asked for over a hundred years.

Two lines of approach can be followed to try to solve the problem of what the angiosperms' ancestors were. One is to compare different fossil groups, looking for possible ancestral angiosperm features. The second is to reconstruct primitive flowers and plants using information from living species. Both methods have certain advantages and if they are used together it is possible to build up a picture of angiosperm evolution.

It is generally agreed that angiosperms must have evolved from gymnosperms which were alive in the early Cretaceous, about 100 million years ago. The main fault here is that the Cretaceous gymnosperms are often thought of as descendants of Jurassic gymnosperms rather than as ancestors of the later Cretaceous angiosperms so scientists have often in the past been looking for different features. The other problem, which can never be overstressed, is that plants disintegrate before they are fossilized. Thus even if a fossil looks like part of a living plant this is no proof at all that the whole plants resembled each other. It always requires a very open mind to prevent preconceived ideas from creeping into comparisons.

There are many Cretaceous gymnosperms but none seems to have the features of an angiosperm ancestor. The cycads, maidenhairs and conifers were developing and diversifying throughout the Cretaceous period but all seem to have been too specialized in their own ways to have given rise to the angiosperms. The cycadeoids developed flowers up to 10cm in diameter. They may have been distinctly coloured but were probably an aid to seed dispersal rather than pollination for they appear to have contained typical gymnosperm ovules which would have been wind pollinated. The seeds themselves may have had fleshy, digestible outer coats.

One other group of plants deserves to be mentioned here because for a long time they were thought to be the 'missing link' in angiosperm ancestry. These were very late seed ferns called *Caytonia* which existed from

the upper Triassic to the lower Cretaceous. They were shrubs whose seeds grew in small fruit-like structures rather like currants. Such 'fruits' have been compared to angiosperm female flower parts, especially because they seemed to have stigmas for receiving pollen. Further investigation into their internal structure has however shown that the so called stigma was a lip near the stalk marking the entrance to a system of channels leading to the ovules. It was there to help support the pollen drop which trapped wind dispersed pollen grains. Pollen was absorbed with the pollen drop into the inner channels where, presumably, sperm was liberated. The ovules themselves, however, were anatomically more like those of angiosperms than of gymnosperms. The new evidence about pollination limited the claim that *Caytonia* was the actual ancestor of angiosperms but it showed that it was nevertheless a very advanced kind of seed fern. It eventually became extinct, possibly because it was unable to compete with the flowering plants.

Early flowering plant fossils

The first undisputed remains of flowering plants have been found in small numbers in early Cretaceous rocks of North America and the USSR. Seeds, leaves, wood and pollen grains are known although no actual flowers have been recorded, possibly because of their delicate

Magnolia flower with (left) a fossil magnolia leaf from the Cretaceous period, found in Labrador. Most so-called primitive flowers really have a mixture of primitive and advanced features, for evolution does not always proceed at the same pace in all parts of a plant. However, magnolias probably have more primitive features than other flowers. They are radially symmetrical and their separate sepals and petals are large, growing from beneath the ovary. They also have many separate carpels and stamens, arranged spirally.

Next page: Fossil fruits, seeds, wood and pollen from various English London clay deposits show the kinds of plants that grew there 48 to 52 million years ago. Shrubs and scrambling plants included *Magnolia* (1), *Hibbertia* (2), *Oncoba* (3), *Uvaria* (4), *Rubus* (5) and *Sabal* (6). Climbers such as *Lygodium* (7), *Vitis* (8), *Iodes* (9) and *Menispermum* (10) twined among trees such as *Pinus* (11), *Platycarya* (12), *Mastixia* (13), *Cinnamomum* (14), *Meliosma* (15), *Dracontomelon* (16), *Corylopsis* (17) and *Nothofagus* (18). Mangroves (*Bruguiera* 19) and *Nipa* (20) grew near brackish muddy water. Ferns probably provided ground cover.

nature. From such sparse beginnings the angiosperms must have steadily increased in importance for the fossil record shows a regular rise in their numbers until they eventually became the most common plant fossils to be found. The apparently sudden appearance of the flowering plants has led many people to suggest that they evolved in upland regions away from places where they were likely to be fossilized. But if we consider the Cretaceous vegetation as a whole we can see that this view is both unnecessary and unscientific.

In the Cretaceous, vegetation varied according to latitude, though there was much less overall variety than there is today. Pollen studies show that there were two main northern coniferous zones with a third zone, more northerly still, where many species related to the maidenhair tree grew. The gymnosperms that lived there were, it seems, adapted for non-tropical habitats. Their leaves were modified to reduce water loss and rings in their wood are the result of seasonal growth.

It is generally assumed that angiosperms evolved in areas where the climate was warm, moist and without seasonal changes. Even in the less varied climate of the Cretaceous, this must mean that they began nearer the Equator and it is in tropical early Cretaceous floras that we should look for their ancestors.

As the Cretaceous continued, the climate became steadily hotter, possibly because of an increase in solar radiation. The rise in temperature would have been imperceptible over the life span of any one plant but would eventually have affected the distribution of the species, allowing colonizing angiosperms to spread both northwards and southwards from their ancestral homes. By the late Cretaceous temperatures were higher than at any time since plants had arrived on dry land and vegetation was less varied over the world than it had been in the late Carboniferous and Permian periods. The higher temperatures may also have stimulated greater growth and speeded up life cycles, perhaps making evolutionary change more rapid.

The gaps in time between the fossil floras we know are considerably longer than the time angiosperms would have needed to spread northwards in these conditions and offer a reasonable explanation for the 'sudden appearance' of the flowering plants.

Primitive angiosperms

One of the ways that flowering plant specialists distinguish less evolved angiosperms is by studying what are called primitive characters. Evolutionary change over millions of years can result in plant organs adapting in a great many ways and to different extents. They are said

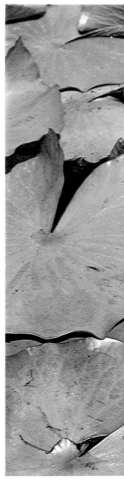

These five flowers all possess some primitive features. They all have free petals, spirally arranged parts, numerous separate stamens and are radially symmetrical and bisexual. All except the rose have their ovary above the sepals and petals and they all occur as solitary flowers. The

Old man's beard, *Clematis vitalba*

Dog rose, *Rosa canina*

carpels of old man's beard and the water lily are still separate although there are fewer of them, while those of the poppy are fused. The rose also has numerous, separate, spirally arranged carpels but these are embedded in a fleshy receptacle which will become the rose hip.

Flowering rush, *Butomus umbellatus*

to be primitive or advanced depending on how great the change has been.

By comparing living plants with one another, botanists have compiled a list of primitive characteristics found in living species. These include such features as woody growth, alternate spacing of the leaves, leaflet-like outgrowths at the base of the leaf stalk (stipules), radially symmetrical flowers with many parts and other anatomical features which can only be seen under a microscope. By examining the whole plant it is therefore possible to obtain an idea of how primitive it is. In the same way genera and families can be compared. No single family has all the primitive characters but some have more than others.

Once primitive families have been identified they can be taken as models that show what early flowers were probably like. Many scientists now accept that the Magnoliaceae is the most primitive living family because it has the greatest number of primitive characters.

Like the Magnoliaceae, the original angiosperms were probably woody shrubs which produced simple radially symmetrical flowers with many separate parts. Their petals and sepals were modified leaves, at first only slightly different from the stem leaves surrounding the flowers. Like today's magnolia flowers, they were probably visited by feeding beetles which ate the juicy parts and accidentally transferred pollen from one flower to another on their body. Beetles today are lured by odours and faint colours such as white and yellow, so there is no reason to suppose that the early flowers were brightly coloured. The fact that they were similar to cycadeoid 'flowers' is very important here. These were probably also visited by beetles so that the animal population was ready for angiosperm flowers and recognized them as a food source. If beetles came to prefer angiosperm flowers as food they may have ceased to visit the cycadeoids and this may have been why the cycadeoids declined and became extinct.

Angiosperm flowers are delicate, short-lived structures so it is not surprising that they are preserved less frequently than other parts of plants. There is as yet no evidence of actual flower parts from early Cretaceous rocks where only a few angiosperm fossils of any kind have been found. The oldest petal comes from middle Cretaceous rocks from Dakota, USA, together with several fruits and flower bases. From these a flower called *Magnolia paleopetala* has been reconstructed. It was perhaps unwise to give it the name of a living genus because we cannot be sure how closely related it was, but it is evidence that magnolia-like flowers were alive about a hundred million years ago.

The supremacy of flowers

The ancestral home of the angiosperms seems to have been in the tropical zone although exactly where is not known. No-one is quite sure how they fitted into the gymnosperm dominated landscape of the time but it seems likely that they probably grew at first outside the main forest areas. Unstable, disturbed environments are much more likely to encourage the rapid and efficient reproductive mechanisms which were evolved by the early angiosperms – because only species which *can* reproduce quickly and efficiently are able to survive and continue to reproduce. In contrast, the dense conifer forests were communities made up of large, steadily growing trees and there was little opportunity for new species relying on short, rapid life spans to invade them.

It has also been suggested that the evolution of the earliest angiosperms was partly a result of the changing eating habits of the Cretaceous dinosaurs. The fossil remains of these giant reptiles suggest that during the early Cretaceous the types that fed mainly on adult trees were becoming extinct while other large, low level browsers were rapidly evolving. Such a change in feeding pattern would have thinned out the conifer forests by destroying the younger plants which had not had time to reach maturity. The faster growing and faster maturing angiosperms would naturally have had a strong competitive advantage under these conditions.

Angiosperms probably developed on the fringes of the conifer forests in areas of loose, rocky, well-drained soil, on slopes where they received irregular supplies of water. Occasional droughts would have provided the impetus for further changes in growth form and anatomy. Variation in leaf shapes, leaf fall, reduction in size and the development of special cells in the wood to transport water more efficiently are obvious adaptations but there were other equally significant but less noticeable changes.

Once the angiosperms had perfected their reproductive systems, adapted their pattern of growth to different conditions and established their associations with pollinating animals, they were able to move into open environments and even into the denser parts of the conifer forests themselves. The many new environments stimulated even more variety as different angiosperms adapted to specialized conditions.

A close look at the types of plants in the tropics today reveals some rather interesting facts: there are far more of some primitive characters in rainforest angiosperms than in plants living elsewhere. This could mean that these are not really primitive characters at all, merely adaptations to the rainforest habitat. However, we know from fossils that most of these same characters were even more abundant in Cretaceous and early Tertiary angiosperms than they are in present day species, so they are certainly very old. The rainforest itself is like a museum of ancient plants, saved from extinction because they are able to grow best in this environment, where they probably first evolved.

The tropics as a whole still contain the largest number of angiosperms and also have the greatest number of species in a given area. Numbers decrease both north and south of the Equator until in the Arctic Circle there are less than a thousand species of flowering plants. Comparison with the far south is difficult because Antarctica is the only large landmass within the Antarctic Circle and it is almost completely covered with ice and snow.

Tropical habitats seem ideal for angiosperms and any move away must have been made possible by evolutionary change either of a structural, reproductive or even physiological nature. The rest of this chapter looks at the kind of changes that occurred, and the reasons for them.

Leaf fall

Tropical plants living in reasonably equable climates grow continuously. Shoots elongate, producing new leaves from their growing tips and shedding the oldest leaves behind them; the trees are evergreens. If such trees try to expand their range into areas where the climate is more variable, they tend to come under stress during the times of the year when the climate is extreme.

Climates that alternate between warm and moist and warm and dry result in too much water loss during the dry seasons; the leaves will wilt and the whole plant may die. An evolutionary adaptation to counteract this is the development of thicker, more woody leaves with thick cuticles. These lose less water and stand up to dessication better. They shed their leaves progressively like tropical plants and are therefore evergreens. Plants of this type are found today in such summer drought areas as the Mediterranean and in warm deserts in various parts of the world.

Climates which alternate between hot and cold, not necessarily between wet and dry, result in rather different adaptations. Here the important thing is to prevent too much water loss from the leaves at times when the roots may be unable to take up water because the soil is frozen and also to prevent frost damage to the leaves themselves. Trees adapted to regions with cold seasons therefore become deciduous: they shed their leaves in the autumn before the cold season starts and grow shoots and leaves again the following spring. The new leaves are preformed before the cold season but remain at an

Fossil sumach leaf from the Oligocene, U.S.A.

Leaf adaptations. 1 Rubber plant. Thick, glossy leaves repel moisture. 2 Horse chestnut. Deciduous leaves come in many shapes. 3 Nettle. Stinging hairs protect against grazers. 4 Clematis. Leaves modified to tendrils for climbing. 5 *Hamadryas.* Hairy leaves protect against cold. 6 Bladderwort and 7 sundew. Modified leaves are insect traps. 8 Cactus. Spiny leaves lessen water loss and reflect light. 9 *Pachyphytum* stores water in fleshy leaves. 10 Thistle. Prickles deter larger grazers. 11 Daffodil bulb. Leaves store food for embryo. 12 Palm and 13 grass leaves grow from base. Grass can withstand being cropped repeatedly.

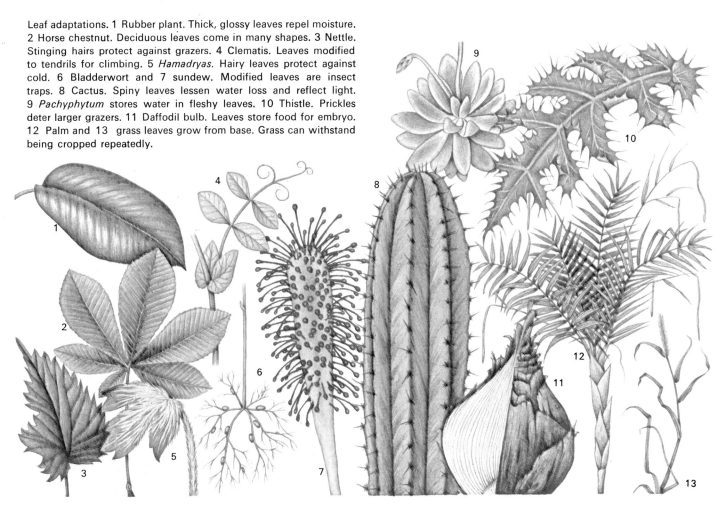

early stage of growth, encased within specially protective bud scales.

There must have been many evolutionary stages on the way to true deciduous plants and the tulip tree (*Liriodendron*) illustrates an intermediate stage. It produces and loses leaves throughout the growing season in the normal tropical way. In the cold season it stops growing and sheds its leaves but it produces no winter buds of preformed leaves. When the warmer weather returns, new leaves grow and are shed throughout the summer. The willows (Salicaceae) show another intermediate stage. They shed all their leaves in the autumn and form winter buds like other deciduous trees. However, not all the leaves for the next year are preformed; after the buds have opened, new leaves continue to grow from shoots in the 'tropical' way. A series of changes of this kind would have led gradually to the evolution of truly deciduous trees.

Conifers had, of course, invaded the temperate regions before the angiosperms evolved. Most survived without losing their leaves annually as their specialized needles are thick, woody and dessication resistant.

Being deciduous is a useful adaptation for life in temperate climates but it is not in itself enough to account for the dominance of angiosperms there. As we shall see, there are many more important reasons for their success.

Herbaceous angiosperms

Herbaceous plants have no lasting parts above ground level. They may be either annuals, living for one year only, or perennials with underground organs which live for many years.

Herbaceous plants have an advantage over ancestral woody ones in some habitats and probably evolved from them by natural selection: the larger, woody plants took longer to reach maturity and in a difficult climate might die before they had produced any seeds. The herbaceous plants, however, brought forward their flowering and shortened their total life spans, reproducing basically at a juvenile stage. Annual plants which flower in their first year and then lose all their aerial parts, are the natural outcome of this evolutionary process. Some annual herbaceous plants such as the goldenrods (*Solidago*) still show traces of their woody ancestors. At the end of the year the aerial shoots start to form wood and look as if they will continue to grow but they then collapse and die instead.

The change from woody to herbaceous types of plants must have occurred many times in different kinds of environment for very many differently adapted herba-

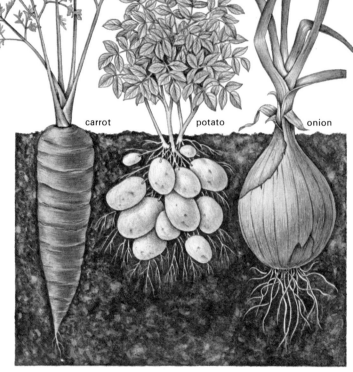

Plants with underground storage organs are able to survive in winter when their green leaves above ground have died down and no photosynthesis can take place.

The pea family is one of the most widespread of all plant families and includes herbaceous and woody species in both tropical and temperate areas.

There is very little fossil evidence for grasses but there is little doubt that they are the world's most successful plants. As they increased in variety late in the Tertiary, there was a parallel increase in the numbers and kinds of grazing animals, some highly adapted to feed on different types of grasses at different stages of growth.

ceous plants now exist. Some families, like the widespread Primulaceae, are mainly herbaceous. This probably means that they became herbaceous early on, before migration and other evolutionary changes within the family. Other families are much more mixed with tropical and temperate, woody and herbaceous species. In these the herbaceous changes probably took place after migration had scattered the species.

As herbaceous plants continued to colonize new areas they became smaller and more and more varied. The pressures of natural selection became even greater on these small, short-lived plants for they had to be successful before their limited life span ended. As a result many of the herbaceous plants that have evolved are highly

specialized for the environments in which they grow. In harsh habitats such as deserts, cold mountain regions and the Arctic tundra, the plants keep their aerial parts as close to the ground as possible to minimize the effects of the climate and grow in rosettes or as small creeping forms. Another way to survive in cold winters is to become partly subterranean, either growing through the soil or developing overwintering parts. These must contain enough food reserves for the dormant period and for the early growth of the plant next spring. Man has naturally taken advantage of many of the larger overwintering organs and now cultivates them for his own use. Among the most familiar are the swollen stems of the potato, swollen roots of the turnip and the food-packed leaves of the onion.

The most successful of all these adapted herbaceous plants must be the grasses, which now cover such vast areas. They were relatively late developing, only appearing in the Tertiary, and had no substantial fossil record until less than 15 million years ago. They now live in such varied habitats that it is difficult to imagine the world without them. They are often the very plants to stabilize the soil, allowing other species to colonize the area. They supply much of the food for grazing animals and also grain for man. Grasses are a long way from the primitive magnolias and clearly show how adaptable the flowering plants can be.

Flowers and their pollinators, including birds, bats, flies, bees and beetles. 1 *Banksia*. 2 Hogweed. 3 *Paphiopedilum* orchid. 4 Red clover. 5 *Cobea scandens*. 6 Sundew. 7 Venus fly trap. 8 Foxglove. 9 Snapdragon. 10 Thistle. 11 *Strelitzia*. 12 *Cobea scandens* after pollination. 13 Carrion plant, *Stapelia*. 14 Water lily. 15 Yucca. 16 Hibiscus. 17 *Ophrys* bee orchid.

Evolutionary change does not always proceed in the same direction and sometimes herbaceous plants changed back to a woody form. Whether or not plants revert depends on two things: their anatomical features and how fixed the herbaceous habit has become in their genetic make up. Woody perennial climbers such as vines and some modern shrubs are probably descended from early herbaceous climbers and shrubs while larger angiosperms such as palms have redeveloped the ancient tree form. The ancestors of other angiosperm trees were probably similarly large, woody plants.

Flowers, animals and pollination

The flower is the main distinguishing feature of the angiosperms. It is by means of the flower that the ovule is fertilized, usually by pollen carried from one plant to another. Fertilization must succeed, or else the species is doomed to extinction. Why, then, are there so many variations of flower form?

The original magnolia type of flower was almost certainly pollinated by beetles, but was equally likely to have provided a source of food for other small crawling and flying animals. Some plants, like *Magnolia* itself, soon developed a defence against the small devouring predators and ensured that ovules and developing seeds were safe against attack. There was little need for the basic flower pattern to change, although small variations would obviously occur. Petal size, number and shape all vary within the family but they remain leaf-like and generally pale in colour. Stamens show much more obvious changes. The earliest were miniature, flattened, leaf-like structures with pollen sacs on either their upper or lower surfaces. Gradually they became smaller and changed to the more typical form with a stalk (the filament) and distinctive pollen sacs (the anthers). All types of stamens are found in the Magnoliaceae family, showing again that they are primitive flowers, still at the experimental stage. The evolution of the carpels which enclose the ovules has already been discussed as one of the key characters of the angiosperms. Living magnolias have completely closed carpels, so they are not entirely primitive in structure; but as we shall see, there are many more advanced types than these.

Evolution from the magnolia type of flower has since produced the enormous variety that we can see today.

As flowers continued to evolve, many became more and more dependent on the animal kingdom for successful reproduction. In return for nectar and pollen as food, animals of all kinds, including birds and bats, transfer pollen from plant to plant. The colours and shapes of flowers have become so specialized that sometimes only one kind of bird or insect can carry out pollination.

Many changed as they became more interdependent with insects other than beetles for pollination. Such flowers developed their own more efficient methods of attracting pollinators. Many have brightly coloured petals and sometimes sepals – colours which are often suited to the vision of particular groups of insects. They are often scented and have nectaries which produce a sweet fluid, giving the pollinators a source of food apart from the actual flower parts and therefore a good reason for visiting the flowers. The carpels have become fewer in number and are fused together into an ovary, often with a single stigma which captures pollen lumps carried on the bodies of visiting insects.

Pollination by larger insects such as bees also brought

about a very noticeable change in flower shape. Many bee flowers are bilaterally symmetrical (like lupins) rather than radially symmetrical (like buttercups). Their shape ensures that the pollinators enter under the stamens and stigma, so that pollen is carried from flower to flower on their back. The orchids provide remarkable examples of adaptation here. The whole stamen is carried on the insect to the next flower where it rubs against the stigma and pollinates it. This type of specialization in shape and structure may occur at first by chance when a mutation produces a 'deformed' flower which may accidentally be more efficient. Its seedlings will inherit its changed shape and will continue to reproduce more successfully than their 'normal' relatives until in time they take over as the typical species form.

Selective pressures have eventually produced many sophisticated flower types which are often pollinated by just one species of insect. In some cases the insect's pollinating role has been taken over by birds and, in a very few flowers, by bats.

Other ancestral flowers developed in rather different ways. Some became specialized for wind pollination, reverting back to the type of pollination found in most gymnosperms, while plants that invaded aquatic habitats occasionally became adapted to pollination by water. Both types developed smaller petals and sepals and sometimes lost them altogether as they would hinder rather than help in these circumstances. Wind pollination evolved in flowers from insect pollination so it is not surprising to find that some species can use either method. Wind pollination has evolved many times in many families but perhaps the most specialized occurs in woody angiosperms with catkins. Most have two kinds of catkins, each with groups of flowers of one sex. There are usually small, rigid catkins with female flowers and longer dangling catkins with male flowers. Male and female catkins sometimes grow on different trees. To help wind transportation of pollen to the female flowers, both catkins are formed early in the year, before the leaves are fully grown. The wind speed is therefore not reduced and no pollen is lost by colliding with the leaves.

Fruits and seeds

The survival of a species depends just as much on the successful establishment of seedlings as on efficient pollination and fertilization. To prevent the parent plants from being overcrowded by new seedlings, the seeds must be dispersed widely and once again the angiosperms seem to excel over the gymnosperms in the

The flowers of both dicotyledons and monocotyledons have evolved into shapes, colours and sizes very different from the early magnolia type. Many are still radially symmetrical (like the buttercup) but others such as sweet pea, iris and orchid are bilaterally symmetrical. Sometimes flowers are grouped together into an inflorescence (grass, bluebell) or compacted into umbels (hogweed) or a compo-

variety of methods they use to achieve this. Some methods are simple but others are more complex, involving wind or even animals.

The primitive form of fruit, the follicle, consisted of a single carpel which dried when it was ripe and opened along one side. The seeds then simply fell out onto the surrounding soil. From such simple beginnings came all the other fruits that exist today.

Fruit evolution was naturally closely connected with the changes that took place in the flower. When the carpels became fused together into many-chambered ovaries compound fruits were produced such as the dry capsules of the poppies, the explosive capsules of touch-me-not (*Impatiens*) and the fleshy tomato berry. Double splitting of a single carpel evolved in the legumes, the group which includes the peas.

Perhaps the most important change which occurred in this stage of plant evolution was when the seeds were retained within permanently enclosing fruits called achenes. This apparently simple alteration occurred many times, giving amazingly successful results. Fleshy

hogweed

paeony

grass

bluebell

iris

poinsettia

orchid

birch catkin

site flower head (sunflower). In natural double flowers the petals are very conspicuous (paeony) but in others they are small and bracts (poinsettia) or a modified style (iris) are the most colourful parts. Grass flowers and catkins are modified for pollination by the wind and have no brightly coloured parts.

fruits such as the single seeded cherry and the many seeded blackberry now rely almost entirely on animals for dispersal, for the fleshy part which forms the animals' food contains substances which prevent the seeds from germinating. These substances are removed when the animal digests the fleshy covering and the seeds pass through with no ill effects. When they are finally ejected, they are some distance from the parent plant and are ready for germination.

Other achenes are dry and are dispersed by the wind, or carried along on the bodies of passing animals. Some are winged (sycamores) some hairy (dandelions) and some hooked (goose-grass). Some dry achenes are also dispersed by animals: the nuts, for example, and false fruits such as strawberries where dry achenes are attached to a swollen central support called a receptacle. The achenes are swallowed with the fleshy fruit.

Another type of achene, the caryopsis, is perhaps the most important of all. This is the grass grain that now supplies much of the world's food in the form of wheat, oats, barley, rye, millet, maize and rice.

Next page: The Sonoran desert of south-western U.S.A. is an area of extremely low rainfall where specialized plants evolved during the Tertiary. These plains today are dominated by creosote bushes (*Larrea tridentata* 1) and bur-sage (*Franseria deltoidea* 2). Other common shrubs are the ocatillo (*Fouqueria splendens* 3), whitethorn (*Acacia constricta* 4) and palo-verde (*Cercidium torreyanum* 5). Cacti are also locally common – saguaro (*Carnegia gigantea* 6), opuntias (*O. engelmanii* 7 and *O. bigelovii* 8, the teddy bear cactus). After the brief rains, small flowering annuals spring up to flower and seed before dying in the following drought.

Continental drift and the evolution of flowering plants

Throughout the evolution of the flowering plants there was a growing interdependence with the animals of the community. Plants made use of feeding animals for reproduction and seed dispersal to an extent that had never occurred before. Gymnosperms and spore-bearing pteridophytes such as ferns continued to reproduce generally in spite rather than because of animal interference. However, the rapid development of the flowering plants at the end of the Cretaceous and beginning of the Tertiary periods was not entirely due to animal interdependence: there were also very strong climatic influences.

Angiosperms were migrating naturally into less equable climates as they spread away from their original tropical home. However, like the conifers, they were also being moved by the break up of the super-continents. The Atlantic Ocean was opening up to such an extent that it became an effective barrier to plant migration during Cretaceous times. North and South America independently moved westwards, pushing up their westerly Rocky Mountains and Andes as they went. They eventually joined up in the Tertiary period although the narrow tropical central American isthmus has not allowed much floral migration in either direction. Australia and Antarctica split away from Africa in the late Cretaceous but remained joined together until well into the Tertiary, about 50 million years ago. Many of the plants that originally lived there were destroyed during this movement while the more southerly movement of Antarctica almost wiped out all plant and animal life there. Australia, with the islands of New Caledonia and New Zealand, ultimately arrived in a tropical zone, while having no tropical plants of its own. When Australia became linked to the Asian landmass about ten million years ago, there was therefore widespread and continuous invasion from the north by species already adapted for tropical climates. These immigrants mixed with the native species to establish many new and local types of plants.

India moved away from Africa towards the end of the Cretaceous period and drifted 5,000km northwards until it collided with the Asian continent and pushed up the Himalayan mountain range. However, as it only moved about 7.5cm a year, most plants were able to adapt as the land passed slowly northwards though the warmer tropical climates and back into the cooler temperatures of the northern hemisphere.

Not surprisingly, continental wanderings had many effects on plant evolution and distribution. Landmasses

Fossil fruit of a plant in the Hamameledaceae family, which includes modern wych hazel. It was found in London clay sediments dating from the Eocene.

A sycamore seed from the Miocene period. Sycamore seeds are dispersed by the wind and may be carried some distance from the parent plant. This fossil was found in the London Clay.

moved into new latitudes and sometimes formed new connections, allowing previously isolated communities to mix. New mountain ranges formed, raising plants into new altitudes and altering local climate by changing wind and rainfall patterns. Some plants could not cope with the altered environments and these died out, but others were able to adapt. One thing seems certain: these dramatic changes were to the advantage of the flowering plants.

In the middle of the Cretaceous there were four main floral zones, simply separated longitudinally, for there was little in the way of high mountains or large seas to upset the continuity. When the continents had moved, however, there were countless different climates, often changing over very short distances. Flowering plants diversified to fill all the areas that could support vascular plant life. This led to some interesting parallel evolutionary developments in widely separated but similar

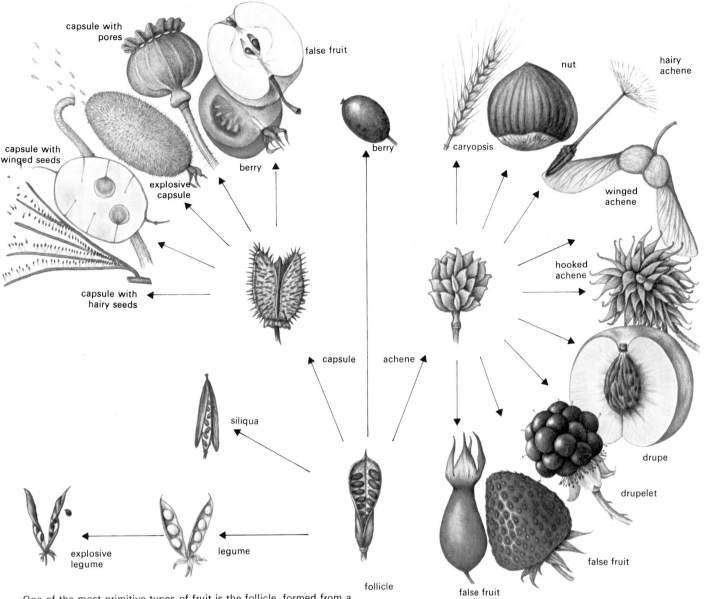

capsule with pores

false fruit

berry

nut

hairy achene

capsule with winged seeds

winged achene

explosive capsule

berry

caryopsis

hooked achene

capsule with hairy seeds

capsule

achene

siliqua

drupe

drupelet

explosive legume

legume

follicle

false fruit

false fruit

One of the most primitive types of fruit is the follicle, formed from a single carpel enclosing several ovules. The seeds were shed when the carpel wall dried and split open. Over millions of years different types of fruits have evolved, adapted to dispersal by mammals, birds, water and the wind.

areas. Adaptations of shape and size to fit a particular type of environment often conceal and confuse family relationships of plants. Some may look superficially like unrelated plants while those that *are* linked may have adapted to different circumstances and come to look very different from one another. Fortunately, flowers are generally not modified by environmental change and they provide reliable standards for evolutionary comparison.

Parallel development of this kind is most noticeable in extreme climates. Cacti developed in the American deserts in the Tertiary period while euphorbias developed in almost exactly the same way in Africa. Both have green, spiny, swollen stems with no leaves and at first sight are identical. However, closer examination reveals that their spines evolved in different ways – and of course their flowers are completely different. Most cacti flowers are large and brightly coloured while

euphorbias are smaller and green with brightly coloured leafy parts often replacing colourful petals, as in poinsettias. Similarly, mangroves all appear very much alike with stems bearing simple leaves and tangled masses of roots. Again, close study proves that there are many different types of mangroves which have evolved in almost identical ways from different families of flowering plants.

Deserts and mangrove swamps are extreme environments where very few gymnosperms and pteridophytes can grow, yet highly specialized angiosperms flourish there. Similarly there are slightly less inhospitable areas such as sand dunes, salt marshes, cliffs, lakes and rivers where angiosperms seem to succeed while other vascular plants fail. There is even the ultimate adaptation shown by the eel grass (*Zostera*) which lives actually in the sea. After 400 million years, land plants have returned to the environment from which they came.

99

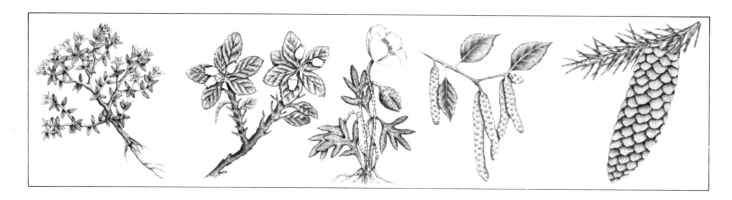

The Ice Ages and after

Flowering plants diversified and developed so much during the Tertiary period that many of the modern groups evolved at this time. The flowering plants also spread throughout most of the world, although there were still some areas that remained dominated by the conifers. During the warmer part of the Tertiary most of the world was covered by a much more luxuriant vegetation than it is today, reflecting the better growing conditions of the time. In Europe and much of southern North America there were vast forest swamps rather like those found today in the Florida Everglades. Great thicknesses of peat were laid down which in time turned into brown coal. South-eastern England had a subtropical flora like that of modern Malaysia, mixed with temperate species. The Arctic regions of Ellesmere Island, Greenland, Iceland and Spitzbergen, were covered with dense conifer forests, with smaller numbers of birch, alder and heathers. Alaska was rather similar but had an even more favourable climate. Birch and alder were much more important plants there and the climate was warm enough for limes, elms and holly to grow.

Towards the end of the Tertiary, important changes were occurring. World temperatures began to drop, forcing many plant species to migrate either towards the Equator or down from the mountains to warmer, low lying areas. As continental drift was also bringing climatic change, some plants had more than one problem to contend with. Nowhere has such a series of

The Ice Ages lasted for some two million years during which the ice advanced and retreated several times. At its greatest extent it covered a large part of Europe, Asia and North America, forcing plants to migrate southwards. Where mountain ranges formed natural barriers to dispersal, whole communities were often wiped out. In other areas the plants survived, migrating north again as the climate improved.

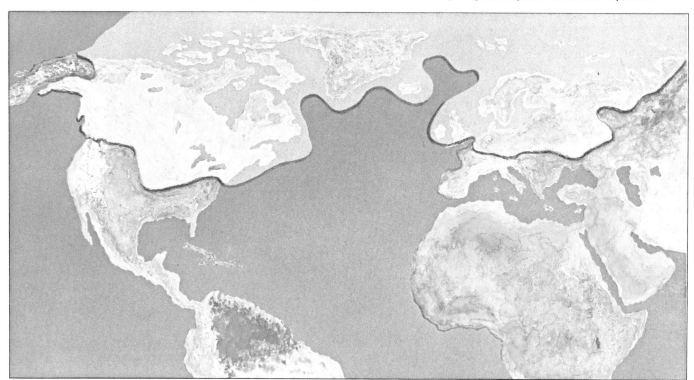

Mosses and lichens colonizing land recently exposed by a retreating glacier.

A community of low-lying plants on the north coast of Spitzbergen: pink flowered *Saxifraga oppositifolia*, yellow *Draba bellii*, and flower spikes of *Oxyria digyna*. A sedge, *Luzula confusa* is behind and a moss, *Rhacomitrium* grows between them.

Spitzbergen is an island within the arctic circle. Today much of it is covered in ice and snow and only a few low lying shrubs, herbs, mosses and lichens grow there.

changes been more fully studied than in the Oregon area of North America. Here there are rich deposits of fossil plants, giving a picture of changing vegetation over some 60 million years. The studies show how the vegetation changed from sub-tropical to mixed temperate forest and then to plants typical of a cooler climate, with an influx of northern species. Further south the climate was becoming much drier and this combination of lower temperature and lower rainfall eventually destroyed many of the Tertiary plants, leaving others growing in much smaller areas.

An even more extreme change was occurring elsewhere for the deserts were just beginning to develop. Ancestors of desert species were slowly evolving in places where there was little rain and, as the general level of humidity and rainfall fell, these semi-arid shrubs, grass and herb communities spread across the new wasteland. Deserts are relatively rare in geological history but today there are many and some are even expanding in area.

Outside the tropical lands, whole communities were destroyed in many areas. In contrast, there were some places where conditions have remained virtually

Next page: Ninety million years ago the climate was much warmer, and there were forests of alder (*Alnus* 1), birch (*Betula* 2) pine (*Pinus* 3), spruces (*Picea* 4), hemlocks (*Tsuga* 5) and dawn redwood (*Metasequoia* 6). The principal shrubs were *Cercidiphyllum* 7, heather (*Calluna* 8) and bog myrtle (*Myrica* 9).

unchanged and here flowering plants have survived unchanged since the Tertiary period. Many islands, the Canaries, for example, kept their wetter climate and still have the type of vegetation that once existed over much larger mainland areas.

Flowering plants in the ice ages

The last million years has been a period of great climatic fluctuation and it is this period which has really shaped the present day distribution of many plants. During this time the average annual temperature of the non-tropical latitudes was alternately warmer and colder than it is now. Lowering of temperatures by only a few degrees increased the amount of snow that fell in winter and reduced melting and evaporation in the summer. This gradually built up the ice at the poles so that ice caps and pack ice spread across the seas. In the mountains, a similar process produced glaciers which began to spread down the valleys. When the temperatures increased, the opposite happened. The polar caps retreated and the mountain glaciers melted and disappeared. There were several fluctuations during the Ice Ages with ice and snow cover varying between about twice what it is today and virtual disappearance. These cold and warm periods are called glacial and interglacial periods.

Plants and animals were forced to migrate in front of the advancing ice sheets but were also able to recolonize the barren wastes left by retreating ice. Whole communities were therefore slowly migrating northwards and southwards throughout this time.

Survival by migration depends on two major factors. Firstly, the species must be able to establish itself by seedling growth in new areas as fast as the parent plants are dying out in the old, less hospitable places. Secondly, there must be suitable habitats available for colonization. Most of the migration disasters and extinctions that occurred during the Ice Ages were caused by lack of habitats. In North America and western Asia the way was open for large scale movements but in Europe the route south was barred for conifers and angiosperms alike by mountain glaciers and by the Mediterranean. Many species perished in Europe because there was nowhere suitable for them to go, leaving the present day flora much poorer than it had been in earlier times.

Pollen analysis

Floral changes in the Ice Ages are investigated by means of pollen grains recovered from deposits formed in the glacial and interglacial periods. In the warmer interglacials peat bogs and fens grew steadily and into these

pollen grains fell from the surrounding vegetation. They also became incorporated into the silt of lake bottoms. When the pollen grains are recovered and compared with pollen from living plants it is possible to identify the plants that once lived in the area. Identification of species is, however, only the start of what can be found out from the pollen grains. Grains taken from different levels of the deposits show how the vegetation was changing and comparisons between different areas show the geographical ranges of different species. By comparing deposits from different glacial or interglacial periods it is possible to work out the gradually changing floras of the Ice Ages.

Plant movements can be monitored by pollen analysis and extinctions can be proved. Certain pioneer plants like birch and pine are often the first trees to colonize new areas, being quickly followed by hazel scrub. In the interglacials the pioneers spread further north, while their original habitats were invaded by oak and elm.

When the last ice age ended some 10,000 years ago, plants gradually migrated northwards to recolonize the areas freed from ice. Analysis of pollen grains in different layers of sediments shows how the vegetation has changed in what is now temperate Europe. Left: Around 10,000 years ago tundra reached almost to the Mediterranean.

Scots pine and birch are the first trees to recolonize ice-free land. By 9,000 years ago they were well established.

5,500 years ago the climate was warm and wet. There were mixed forests of oak, lime and elm with alder in wetter areas.

3,000 years ago the climate was still warm but drier. Oak, beech and hornbeam were the dominant trees.

Today oak and beech are still common in lowland areas but natural vegetation has been greatly modified by man.

Dense woodlands developed, excluding the early pioneers but including species such as lime and alder which preferred the warmer climate. The returning cold reversed the direction of plant movements. Species migrated southwards in reverse order, with the Arctic tundra plants being the last to arrive before the advancing ice. The order of community migration seen in the pollen records can be seen today in the north-south line up of plants, poised as though ready for movement in either direction.

During full glacial conditions, the distribution of cold-loving plants altered in quite a different way from that of warm-loving plants which simply moved as far south as they could. The cold brought plants southwards from the Arctic into isolated mountainous regions. When the temperatures rose again some cold-loving plants moved northwards but others moved up the mountains into the colder, higher altitudes. This was the beginning of the separated distributions of Arctic-

alpines now so popular with gardeners.

By analysing pollen found in peat formed since the last glacial period it is possible to follow changes in vegetation right through to the present day. The last 10,000 years, known as the post-glacial period, is rather like an interglacial. Plant invasions have occurred and have stopped at a position similar to that found in the middle of an interglacial period. However, there are no signs that the climate will deteriorate again and alter the potential distribution of plants. We say potential distribution here because of one important fact. In the last 10,000 years mankind has increased dramatically and has now brought about the wholesale destruction of natural vegetation and introduced his own agricultural plants. Man now becomes a key figure: his agricultural and aesthetic needs have led him to develop better and better crops and ornamental garden plants and it is to this controlled evolution that we must now turn.

Plant evolution and man

Plants had been slowly evolving and adapting to a changing world for 450 million years. Then, a mere 10,000 years ago a new and major influence appeared. Man, who had lived as a hunter gatherer began to take a few plants and animals under his control.

No doubt his first cultivations were entirely accidental, as seeds were dropped and lost around his dwelling places. Hygiene was unknown and the piles of refuse and dung, rich in nitrogenous waste, would have stimulated seed germination and provided ideal conditions for growth. This then could have easily suggested to early man the idea of deliberately using some of his collected seeds to grow plants close to his home, giving an easily and readily available supply of food.

The first attempts would have been crude but as time went on he learned the effects of soil, sun and water on plant growth and soil clearance, tilling and watering crept into his way of life. By such simple means the early farmers learned to exploit the plants that would eventually become their major food source.

The yield of the plant was naturally of prime importance but it was many thousands of years before conscious efforts were made to find ways of improving it. The earliest and perhaps simplest way is to select the best plants and to grow the next generation from their seeds. If this is done for many successful generations, the natural range of variability moves towards the type of plants desired by man. By this type of selection we have

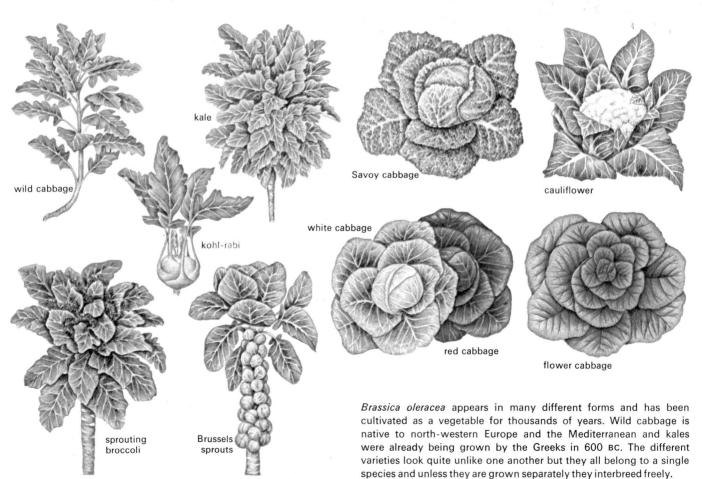

wild cabbage · kale · kohl-rabi · sprouting broccoli · Brussels sprouts · Savoy cabbage · white cabbage · cauliflower · red cabbage · flower cabbage

Brassica oleracea appears in many different forms and has been cultivated as a vegetable for thousands of years. Wild cabbage is native to north-western Europe and the Mediterranean and kales were already being grown by the Greeks in 600 BC. The different varieties look quite unlike one another but they all belong to a single species and unless they are grown separately they interbreed freely.

improved plants such as barley, rice, maize, flax, tomatoes, soya beans, sugarbeet and the many 'double' flowers that have been brought into cultivation.

The development of sugarbeet is perhaps the best documented of all these, since it occurred so recently. Both sugarbeet and the mangel wurzel originated from the seabeet which grows around much of the European coast. Its fleshy root contains both starch and sugar as winter food reserves. The mangel wurzel has been grown

Most main wheat species today are polyploids. One of the first known wheats was wild einkorn (a diploid) which hybridized naturally with the diploid grass *Aegilops speltoides* to produce tetraploid wild emmer. Both these wheats still grow wild in the Near East. Wild emmer was cultivated by early farmers and by 8,000 BC had developed into emmer. By 2,500 BC emmer had crossed naturally with a diploid wild grass *Ae. squarrosa* which grew as a weed among the crops. The result of this cross was one of the bread wheats, spelt, which has a stronger seed stalk and grains that are more easliy threshed.

Two other important species are the tetraploid durum wheat, a further developed form of tetraploid emmer: and diploid einkorn, which has evolved from the original diploid wild einkorn. Durum wheat has a high gluten content and is used for pasta. Einkorn is used as a whole grain animal food.

as animal food for hundreds of years, but the main exploitation of seabeet took place because of its sugar content. The first result of bringing the seabeet into cultivation was the beetroot which became popular as a sweet vegetable. Selection of the sweetest varieties was then started in a much more serious way as a result of the Napoleonic war at the beginning of the nineteenth century. British coastal blockades prevented imported sugar cane from the West Indies from reaching the continental ports, so breeders selected and improved their home grown white beet crops as an alternative sugar source. The sugar content increased from about 2 to 10%, making beet a much more desirable plant which quickly became widespread and important. Further selection has increased this sugar content even more so that today's cultivated sugar beet has a sugar content of nearly 20%.

Sometimes natural variation within a species produces more than one economically useful plant. For example the wild cabbage (*Brassica oleracea*) has given rise to plain, wrinkled and red leaved cabbages, brussels sprouts, cauliflowers, kohl-rabi and kale. One disadvantage of their common ancestry is that they will

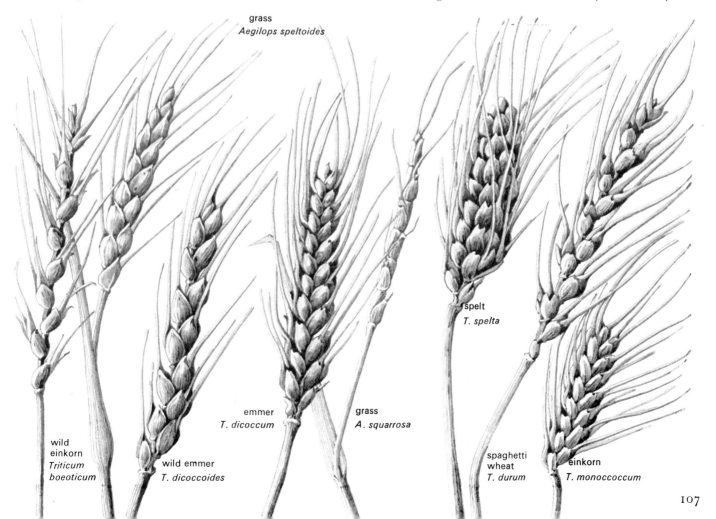

grass
Aegilops speltoides

emmer
T. dicoccum

grass
A. squarrosa

spelt
T. spelta

wild einkorn
Triticum boeoticum

wild emmer
T. dicoccoides

spaghetti wheat
T. durum

einkorn
T. monoccoccum

all interbreed so must all be grown separately to maintain their individual features. Fortunately this is not a real problem today for economical cultivation relies on growing large numbers of one type of plant.

Hybridization

Selection alone is insufficient to explain the existence of the vast majority of species that have arisen naturally or within cultivation. There is also the process of hybridization by which two plants from different races, species or even genera may interbreed to produce offspring. Hybridization was not really understood until less than a hundred years ago although it has been utilized in plant cultivation for thousands of years. Man, the farmer, was migrating from the Near East and, as these early settlers spread around the world, they took their crops with them. Plants were thus being deliberately removed from their native areas and natural habitats and taken into many new parts of the world. Wheat was foremost among these early crops, although it was of a very primitive form. As man moved, his cultivated wheat hybridized with certain local grasses, producing changes for the better. Not only was the yield improved, but the grains also became held more firmly within the wheat head so that much less grain was lost before and during harvesting. These changes may have been noticed by the farmers but they would certainly not have been understood. Man was merely selecting the most desirable plants from the hybrids that were occurring naturally as he slowly migrated across Europe. The result was the appearance of several different strains of wheat suitable for the varying needs of the scattered human communities.

Of course, now that the process of hybridization is understood, man has taken advantage of his knowledge to produce new and important varieties by controlled experimental cross-breedings. These must be carefully tested before the value of the new varieties can be established and in fact many new types have proved unsuitable as crops. Others are successful and modern farmers can choose from a large range of wheat varieties to suit the needs of different agricultural and climatic conditions around the world.

Polyploidy

There is a third process that can alter plants and has led to the development or improvement of many domesticated crops. It is the result of an increase in the amount of nuclear material within the plant's cells. This is called polyploidy and, at its simplest, involves the natural doubling of the usual number of chromosomes

The origins of Hybrid perpetual roses
Hybrid perpetuals were the result of crosses between descendants of two European species (*Rosa damascena* 1 and *R. gallica* 2) and of two Chinese (*R. chinensis* 3 and *R. gigantea* 4). All these had been cultivated for hundreds of years but of the European roses only one group, the Autumn damasks, flowered more than once a year. Slater's crimson china (5) a perpetual flowering variety, was crossed with one of the European roses to produce the first Portland rose (6) in 1800. If properly pruned, these would flower more than once. Rose du roi (7) was bred from the early Portlands in 1816 and is often called the first Hybrid perpetual: it flowered continuously with no special treatment. Meanwhile another Chinese rose, Hume's blush tea scented china (8) had been crossed with *R. gallica* to produce Hybrid chinas (9). Crosses between some of these and Rose du roi produced larger flowered Portlands (10) and further interbreeding between Portlands, Hybrid chinas and Bourbon shrub roses (11) led in 1837 to the larger flowered Hybrid perpetuals (12) which were to remain popular for over half a century. They were then largely replaced by their own descendants, the Hybrid teas.

that occur in the species. It is certainly not a freak process and is of genuine relevance to plant evolution and breeding. At least a third of all domesticated species and nearly three-quarters of forage grasses are estimated to be polyploids.

The reason for this spontaneous increase in chromosome number is not fully understood but the results are clearly appreciated and often exploited by man. Spontaneous polyploidy in a plant often gives it larger flowers and fruits – a very desirable feature in any domesticated plant. Present day apples, pears, potatoes and peanuts are all the result of polyploidy. All have much larger food storage organs than their wild ancestors and all are much more valuable as cultivated crops. Similarly the larger, showy and more beautiful flower heads of certain polyploids have made them popular as ornamental plants.

This increase in genetic material is often linked with hybridization. Frequently hybrids are sterile and must eventually die without leaving any offspring. But occasionally polyploidy provides the means for the hybrids to become fertile, ensuring a succeeding generation. If these new hybrids can then compete successfully with their parental forms and survive long enough to produce sufficient descendants, their future is assured. If not they will be doomed to extinction. Most hybrids and polyploids do not survive, but there are many successful examples that have been exploited by man. By these processes, we now have not only wheat but also oats, sugar cane and tobacco.

Polyploidy seems an easy, rapid and relatively efficient means of improving plants and with the resources available nowadays for inducing hybridization and polyploidy some success has been achieved. The problem is that only a minute fraction of the plants that reach maturity have actually improved in the way the breeder requires. Success in producing suitable varieties probably lies in using a combination of induced genetic change and intensive selection. Clearly much work still remains to be done before this is universally successful.

Current aims of plant breeding

It should now be apparent that there are great advantages in plant breeding and that the common aim is either to increase the yield of crop plants or to produce more showy ornamentals. Crop plants will be the eventual key to man's survival, especially as half the world's population is undernourished or starving. The amount of land available for cultivation is limited so it is plant yield that is of prime importance.

Yield can be measured as the amount of food crop

cocksfoot
Dactylis glomerata

groundnut
Arachis hypogaea

apple
Malus pumila

Many fruits and vegetables are the result of polyploidy, which often leads to larger leaves, flowers or fruits. Polyploids are often infertile and although experiments are still being made, artificially induced polyploidy does not seem to provide an easy answer to the world's food problems. The plants shown here are mainly wild polyploids that have been commercially selected. Triticale is a cross between wheat and rye – the first successful attempt to breed from different genera. It is hardy like rye but has a large grain spike and fuller kernels, more like wheat.

produced per plant or per area of cultivation and it is dependent on many factors. As we have seen, all plants vary naturally to some extent but external influences also play a large part in controlling the ultimate yield. Different plants are affected in different ways by temperature, sunlight, rainfall and soil type and it is important to grow the right crop in the right environment. Disease can also affect yields quite disastrously at times but again, its impact is variable and one crop may be affected more than another.

The problem is therefore a complex one. The breeder must produce new forms with higher yields and with a greater tolerance to difficult climatic conditions and

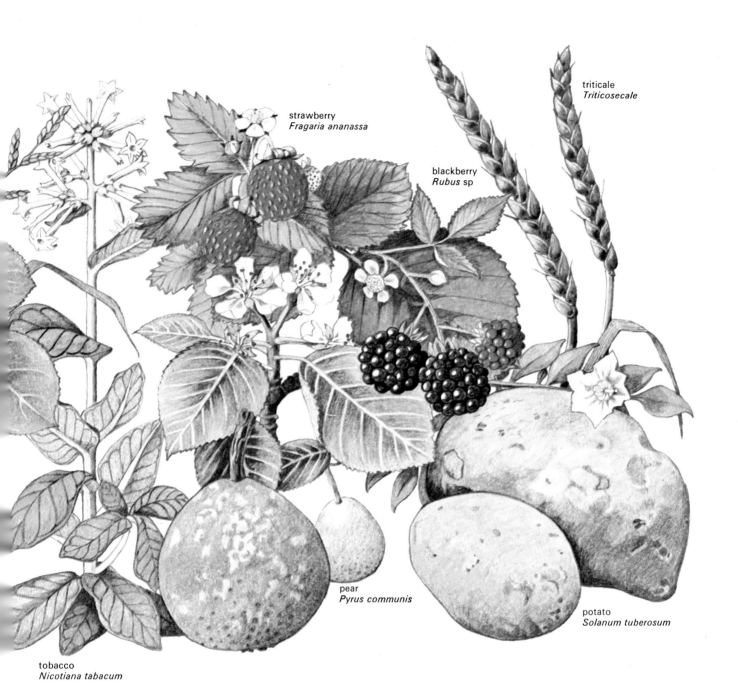

strawberry
Fragaria ananassa

triticale
Triticosecale

blackberry
Rubus sp

pear
Pyrus communis

potato
Solanum tuberosum

tobacco
Nicotiana tabacum

greater resistance to disease. It does not matter to him if certain characters vital to the plant's survival in the wild are lost, for they may no longer be necessary under cultivation. The grower's aim is to increase productivity. Many plants can no longer produce another generation without the help of man: seed dispersal may be impossible as in the case of maize or delayed, as in many forage grasses. Life spans have been altered to meet certain climatic conditions, to allow harvesting at different times or even to give more than one crop per year. In addition to these 'productivity' modifications, breeders often try to make plants uniform in size and shape so that harvesting is easier and more predictable.

If the crop all matures at the same time and does not need much grading, it is a great advantage, especially now that modern farming is highly mechanized.

Plant breeding is a highly specialized and complex science but it is really quite new. The mechanism of fertilization itself has only been understood for a little more than a hundred years while truly scientific breeding did not begin until about sixty years ago. There is no doubt that the demand for plant breeding will increase. It is by understanding and exploiting such controlled evolutionary changes that man will be able to survive and improve the living standards of the whole world population.

The future

Evolution seems obvious when we look back in time at the changing patterns of plant form and reproductive methods. But it is not so evident if we look only at living plants or to the future. There is no reason to suppose that living plants are incapable of further evolution and indeed plant breeding shows how change is possible. Providing man does not interfere too much in too many ways, plants will certainly continue to evolve.

Unfortunately there is a very real danger of far too many species becoming extinct before they have even been found and named. Simultaneously, modern methods of cultivation are developing uniformity in crops, destroying their natural variability. A few missing species may be unimportant but wholesale destruction in many areas could be disastrous both for natural evolution and for future breeding programmes.

Crops are not of course the only plants important to man. It is all too easy to upset the delicate balance of the natural world in which animals and plants are so closely interrelated. Indiscriminate use of weedkiller may destroy insects and birds; clearing forests may permanently affect rainfall patterns; overintensive agriculture may turn fertile land into desert.

The problem has been recognized but it is a difficult one to solve. There are two apparent ways of preventing plants from becoming extinct. One is to conserve enough habitats in which the plants can develop naturally; the other is to remove plants or their seeds to collections. Both methods have pitfalls and can never be entirely successful, especially for the larger plants. For natural evolution to continue it has been estimated that from 1,000 to 25,000 plants of one type must be living together and for large trees, which may not be fully mature until they are many metres tall, a very large area indeed is necessary. More and more land is needed for agriculture so continuous effort is needed to conserve enough habitats without altering them by management programmes designed to maintain them.

Plant collections preserve and cultivate different varieties by artificial means and nearly half a million strains and varieties are now in various collections around the world. The next obvious step is to establish specialist collection centres in areas where potentially important plants naturally grow. This, combined with some conservation within the same areas would preserve species and provide experts to work on them.

In a world that is always needing more food, plant improvement is desperately needed. By understanding how and why plants have evolved in the past, we can, we hope, learn to conserve them in sufficient numbers to allow the slow processes of evolution to continue.

Modern farming methods are causing wild flowers such as this fritillary to become rare. In North Meadow, Wiltshire, the species is being preserved by haymaking in the traditional way, excluding grazing animals during the growing and flowering season.

Below: An experimental fruit farm in the Negev desert. With efficient irrigation and use of fertilizers, the dry sand of the desert can be turned into workable farm land.

Many plants contain valuable medicinal compounds: Digitalin from foxgloves is used in the treatment of heart disease. Many other substances may be obtainable from a wide range of plants that have not yet been studied and there is a risk that some will become extinct before their value can be assessed.

Above: A mixed plantation of beech and pine. The faster growing pine will be felled first, leaving the beech to mature more slowly. Well planned forestry provides a renewable resource and can preserve habitats for both plants and animals.

Below: Amazon forests in South America cover 5 million square kilometres, releasing enormous amounts of oxygen into the atmosphere. The long-term effects of destroying rainforest for timber, agriculture and ranching, could be disastrous.

The paper industry uses millions of trees each year, mainly for newsprint. Recycling wastepaper and new methods of paper-making help to preserve timber stocks. This book is printed on wood-free paper.

Bibliography

ANDREWS, H N: *Studies in Paleobotany*, John Wiley, New York, 1961

ASHTON, B G: *Genes, Chromosomes and Evolution. The Principles of Modern Biology*, Longman, London, 1967

BANKS, H P: *Evolution and Plants of the Past*. Fundamentals of Botany series. Macmillan, London, 1972.

BELL, P AND WOODCOCK, C: *The Diversity of Green Plants* (2ed), Edward Arnold, London, 1972

BIERHORST, D W: *Morphology of Vascular Plants*, Macmillan, New York; Collier-Macmillan, London, 1971

BOLD, H C AND WYNNE, M J: *Introduction to the Algae and Reproduction*, Prentice Hall, Inc., New Jersey, 1978

BRIGGS, D AND WALTERS, S M: *Plant Variation and Evolution*, World University Library, Weidenfeld and Nicholson, London, 1969

DELEVORYAS, T: *Plant Diversification* (2ed). Modern Biology Series, Holt, Rinehart and Winston, New York, London, Sydney, 1977

HUGHES, N F: *Palaeobiology of Angiosperm Origins*, Cambridge University Press, Cambridge, 1976

INGOLD, C T: *The Biology of the Fungi* (3ed), Hutchinson Biological Monographs, Hutchinson, London, 1976

MARTIN, E A: *A dictionary of Life Sciences*, Macmillan, New York, 1976; Pan Books, London, 1978

ROUND, F E: *The Biology of the Algae*, Edward Arnold, London, 1965

SPORNE, K R: *The Mysterious Origin of Flowering Plants*, Oxford Biology Readers No. 3, Oxford University Press, Oxford, 1971

SIMMONDS, N W: *Evolution of Crop Plants*, Longman, London and New York, 1976

TAKHTAJAN, A: *Flowering Plants, Origin and Dispersal*, Oliver and Boyd, Edinburgh, 1969

WATSON, E V: *The Structure and Life of Bryophytes* (2ed), Hutchinson University Library, Hutchinson, London, 1967

WEBSTER, J: *Introduction to Fungi*, Cambridge University Press, Cambridge, 190

WEST, R G: *Studying the Past by Pollen Analysis*, Oxford Biology Readers No. 10, Oxford University Press, Oxford, 1971

Glossary

ANGIOSPERM Vascular plants with seeds in ovaries; the flowering plants.

ARCHEGONIUM Simple female sex organ of plants where egg cells are produced.

BRYOPHYTES Mosses, hornworts and liverworts – simple land plants without vascular tissue or roots.

CARPEL Leaf-like structure that encloses the ovule, sometimes individual (pea pod), sometimes fused together into ovaries.

CHROMOSOME Thread-like structures of cells that carry the genes.

DNA (Deoxyribonucleic acid) Main constituent of chromosomes, carrying the genetic code.

EMBRYO The structure formed after the egg cell has been fertilized. It grows into the new plant.

EPIDERMIS Outermost layer of cells of leaves and stems.

FLOWER The sexual reproductive structure of angiosperms.

FOSSIL Preserved evidence of living things from the past.

GAMETES Specialized reproductive cells that fuse in pairs to reproduce the next generation of plants.

GAMETOPHYTE Plant generation that has sex organs and produces gametes.

GENE Particle contained in chromosomes that determines inherited characteristics.

GENUS (pl. genera) A group of living organisms divided in turn into species.

GYMNOSPERM Vascular plants with ovules not enclosed in ovaries, e.g. conifers.

HETEROSPOROUS Producing two kinds of spores. The spores mature into plants having either all male or all female reproductive cells.

HOMOSPOROUS Producing only one kind of spore. Spores grow into plants with both male and female reproductive cells.

HYBRID Plants (or animals) produced by crossing two different species or varieties, occasionally from two different genera.

INTEGUMENT Protective outer layer of the ovule of seed plants.

NUCELLUS Tissue inside the outer integument of the ovules of

gymnosperms and angiosperms, corresponding to the sporangium in other plants.

PALYNOLOGY The study of fossil spores and pollen grains.

PHOTOSYNTHESIS The process by which green plants use the energy of sunlight to convert water and carbon dioxide to sugars. Oxygen is produced as a by-product.

PROTHALLUS The gametophyte (sexual) plant of ferns and other spore-bearing plants and the equivalent stage in gymnosperms.

PTERIDOPHYTES Members of one of the main divisions of the plant kingdom – vascular plants with two generations of different plants, one reproducing sexually by gametes, the other asexually by spores. Includes ferns, horsetails and clubmosses.

OVARY Female part of flower, containing the ovules.

OVULE Female reproductive organ in gymnosperms and angiosperms that matures into seed when fertilized.

PHLOEM System by which food manufactured in the leaves is distributed throughout a plant.

SEED Formed from a fertilized ovule, contains the embryo.

SPECIES A distinct group of plants (or animals) that can successfully interbreed, a sub-division of genus.

SPORANGIUM The sac in which spores are produced.

SPORE Small, usually single-celled reproductive body produced by sporophyte generation of plants such as ferns, horsetails and mosses. Grows to become new sporophyte plant.

SPOROPHYTE Plant generation producing spores.

STOMA (pl. stomata) Small openings in epidermis of plants through which gases enter and leave the plant.

VASCULAR PLANTS Plants containing vascular tissue, the conducting system which enables water and minerals to pass throughout the plant.

XYLEM System in which water and minerals travel through a plant from the roots.

ZYGOTE Cell produced when two gametes fuse – the first cell of the new embryo plant.

Index

Acknowledgements

In preparing the reconstructions the following references have been used:
page 21 D. A. Eggert (*Lepidodendron*); 33 K. J. Niklas, T. C. Phillips (*Protosalvinia*), D. Edwards (*Cooksonia*); 34 R. Kräusel, H. Weyland (*Drepanophycus*), R. Kidston, W. H. Lang (*Asteroxylon, Rhynia*), J. Walton (*Zosterophyllum*), A. E. Kasper, H. N. Andrews (*Pertica*), H. P. Banks, S. Leclercq, F. M. Hueber (*Psilophyton*); 42 K. Magdefrau (*Pleuromeia, Nathorstiana*); 43 M. Hirmer (*Sigillaria*); 44 M. Hirmer (*Calamites*), R. von Wettstein (*Sphenophyllum*); 48 R. Kräusel, H. Weyland (*Cladoxylon*), E. W. Binney (*Stauropteris*), J. Morgan (*Psaronius*), B. Thomas (*Osmundites kolbei*), H. N. Andrews and Kern (*Tempskya*); 53 Charles B. Beck (*Archaeopteris*), P. M. Bonamo, H. P. Banks (*Tetraxylopteris*); 55 W. N. Stewart, T. Delevoryas (*Medullosa*), T. M. Harris, H. Thomas (*Caytonia*); 59 C. Grand'Eury, D. H. Scott (*Dorycordaites*); A. A. Cridland (*Amelyon*); 66 W. Gotham, H. Weyland (*Palaeocycas*), B. Sahni (*Williamsonia*), M. Hirmer (*Cycadeoidea*); 76 C. Grand'Eury (*Cordaianthus*), T. Delevoryas (*Lebachia*). Reference for the spread on pp. 84–5 was provided by Margaret Collinson.

Artists
Rudolph Britto 14–15, 20, 23, 34, 40, 44, 64, 65, 70
Sally Caldecott 92–3, 107, 108–9, 110–11
Diagram Group 13
David Etchell/John Ridyard 12, 18–19, 22, 25, 28–9, 39, 41, 42, 50, 52–3, 54, 61, 68–9, 70–1, 80–1, 106
Will Giles 20, 21, 31, 32–3, 43, 53, 55, 66–7, 89, 90, 99, 106
Sarah Hartley Edwards 24, 32, 42, 52, 65, 80, 88, 106
Ingrid Jacob 48–9
Lesley McKinnon 13, 94–5
Tony Swift 36–7, 46–7, 56–7, 62–3, 72–3, 84–5, 96–7, 102–3
Shirley Tuckley 34–5, 44–5, 58–9

Photographers
A–Z Botanical collection 41, 48, 66, 113
Heather Angel 14, 15, 16, 26, 27, 30, 31, 35, 40, 41, 43, 51, 68, 74, 75, 77, 78, 79, 83, 86–7
Donald Baird 45, 59, 68, 71, 83
British Museum (Natural History) 17, 38, 43, 44, 45, 49, 53, 55, 60, 67, 81, 88, 98
Fred Bruemmer 104–5
Brooks/Bruce Coleman Ltd 101
William Chaloner 18
Bruce Coleman 78
David Dilcher 17
Robin Fletcher 112, 113
Michael Fogden 6–7, 65, 77
Imitor 18, 44, 65
Eric Kay 101
Oxford Scientific Films 8, 10, 14, 26, 30, 50, 76, 91
William Paton 43
E. Ross 113
Harry Smith 113
D. Spicer 101

The publishers wish to thank Barbara Greenwood and Philip Phillips (Merseyside County Museums), Jim Keesing, and P. R. Crane for their help.

Index by M. O'Hanlon

Designed by Marion Neville

GRAFICAS REUNIDAS, S. A.
Av. de Aragón, 56 —Madrid-27.